Advanced Simulation of Alternative Energy

Advanced Simulation of Alternative Energy

Simulation with Simulink® and SimPowerSystems™

VIKTOR M. PERELMUTER

CRC Press
Taylor & Francis Group
Boca Raton London New York

CRC Press is an imprint of the
Taylor & Francis Group, an **informa** business

MATLAB®/Simulink® is a trademark of The MathWorks, Inc. and is used with permission. The MathWorks does not warrant the accuracy of the text or exercises in this book. This book's use or discussion of MATLAB®/Simulink® software or related products does not constitute endorsement or sponsorship by The MathWorks of a particular pedagogical approach or particular use of the MATLAB®/Simulink® software.

CRC Press
Taylor & Francis Group
6000 Broken Sound Parkway NW, Suite 300
Boca Raton, FL 33487-2742

First issued in paperback 2022

© 2020 by Taylor & Francis Group, LLC
CRC Press is an imprint of Taylor & Francis Group, an Informa business

No claim to original U.S. Government works

ISBN-13: 978-0-367-33957-9 (hbk)
ISBN-13: 978-1-03-233655-8 (pbk)
DOI: 10.1201/9780429324055

Visit the Taylor & Francis Web site at
http://www.taylorandfrancis.com

and the CRC Press Web site at
http://www.crcpress.com

Library of Congress Cataloging-in-Publication Data

ISBN: 978-0-367-33957-9 (hbk)
ISBN: 978-0-429-32405-5 (ebk)

Visit the eResources: www.crcpress.com/Advanced-Simulation-of-Alternative-Energy-Simulation-with-Simulink-and/Perelmuter/p/book/9780367339579

Dedicated to my family, friends, and colleagues.

Contents

Preface

Renewable sources of the electrical energy (RES) today are the mainstream in electrical engineering. A lot of engineers, scientists, and practitioners are engaged in the development, research, commissioning, and maintenance of RESs. The availability of modern and effective methods of mathematical modeling of such systems can significantly simplify and accelerate the execution of these works.

Creating the modern RES is a multidisciplinary task, which requires the involvement of specialists from various branches of knowledge. However, in general, the creation of modern RES is impossible without utilization of the latest achievements of electrical engineering, first of all, systems and methods of power electronics.

The set of SimPowerSystems™ blocks, which is included in the MATLAB®/ Simulink® programming environment, is intended to simulate various electrical systems: industrial electronics, electrical machines, electrical drives, production and distribution of the electric energy, etc. The use of SimPowerSystems greatly facilitates the creation of models of electrical objects. On its basis, as well as using Simulink® blocks, it is possible to create models of new, perspective electrical devices, which are under development. All this causes the popularity of models Simulink/SimPowerSystems. It should be noted that the name of this set of blocks changed in different versions of MATLAB®, and in R2019 it is called Electrical/Specialized Power Systems, but we will keep the former name for succession.

There are several books on RES simulation issues. In 2016, CRC Press published the book of the author, V. Perelmuter, *Renewable Energy Systems; Simulation with Simulink® and SimPowerSystems™*, in which the problems of simulation of various wind generators (WGs), as well as systems with batteries, photovoltaic (PV) systems, the systems with fuel cells (FCs), microturbines, and some hydroelectric systems were considered. The peculiarity of this book is that about 150 models of renewable energy sources were appended to the book, which were developed with the use of SimPowerSystems. The aims of these models are as follows: to help in studying SimPowerSystems; to help in studying the various electrical engineering fields, such as industrial electronics, electrical machines, electrical drives, production and distribution of the electrical energy, etc.; to facilitate understanding of various renewable energy system functions; and to help readers as a basis for developing their own systems in these fields.

But since the renewable sources of the electric energy are, as before, in the center of the scientific investigations and thousands of the articles and conference papers appear every year, many of which contain very interesting ideas and industrial elaborations, it seems appropriate to continue work on creating models of new, promising RES with the above-mentioned objectives, as well as to acquaint readers with new trends in the considered technical field. This book has a goal to perform this task. It considers the modeling of the same RES (WG, PV array, FC, and microturbine) as in the previous book, but using new models of devices and systems developed by the author. For a number of systems, more specific details of their work are simulated. The part concerning the use of water sources of energy is significantly elaborated. The models of the hybrid systems containing sources of electrical energy of different types (microgrid) are added.

In fact, this book is a sequel of the previous one and contains many references to it to save space. More than 100 models of quite complex RESs of the above-mentioned types are appended to it. These models are found at https://www.crcpress.com/9780367339579. The author tried to avoid complex models that are of interest only to a small number of specialists in this field, as well as models containing new, not yet proofed proposals and improvements whose prospects are not clear; only the main senses and trends are reflected in the models of the presented systems.

During the work with the aforementioned models, the reader can investigate them with the given or the chosen parameters, use them as a basis for the analogous models developed by him or her and so on. Interactive work with a computer is supposed to be done during reading the book.

The content of this book can be summarized as follows:

Chapter 1 presents a brief description of the SimPowerSystems blocks that are mostly used in the RES models in this book. In the first part of the chapter, the standard blocks from SimPowerSystems library are described, including electrical sources, loads and transformers, transmission lines, power electronic devices and circuits, and electric generators. In the second part, more complicated models of devices and units are described. These units are developed partly in SimPowerSystems and partly by the author using SimPowerSystems blocks, including batteries, supercapacitors, multilevel power converters, and multiphase electrical generators.

Chapter 2 considers the advanced models of the WG plants, which are the further development of models discussed in the earlier books. The main relationships describing WG operation and also the methods for WG operation optimization are given. Such special features such as tower shadow effect, active damping, and system inertia support are under consideration. Utilization of the multiphase generators is very promising in the WG of large power; such a WG with six- and nine-phase induction and synchronous generators are simulated in this chapter. Direct current transmission systems with the high voltage are very perspective for offshore wind park, and the series-connected WG gives an opportunity dispensing with bulky and costly offshore substations. Simulation of these systems with different onshore inverters, as well as two-terminal offshore wind parks, is under consideration.

Chapter 3 covers simulation of both grid-connected and island operating photovoltaic (PV) systems. Both developed by the author and included in SimPowerSystems models of PV cells are used for building models of PV arrays. Power rise of the PV energy sources promotes the employment of the multilevel inverters; thereupon, utilization of the cascaded H-bridge multilevel inverters is very promising, and their simulation is considered in the chapter when the various variants of the system structure are used. At this, the methods to decrease interphase and per-phase unbalances are investigated. For islanding operation, the methods to identify this mode are simulated. Parallel operation of two PVs with isolated load under different environments for PVs is simulated; the droop characteristics are used to provide parallel work.

Fuel cells and microturbines that are able to produce electricity regardless of weather conditions are considered in Chapter 4. The polymer electrolyte membrane fuel cell (PEMFC) model developed in this chapter makes it possible to simulate approximately the processes of stopping and starting a fuel cell and incorporates the model of the gas feed mechanism. A supercapacitor is used to power the load during short breaks in the operation of fuel cells. To obtain hydrogen, an electrolyzer is used, the model of which is developed in this chapter. A model of high-temperature solid oxide fuel cells (SOFCs) has also been developed, which is preferred for use in stationary installations as well as in combination with microturbines, forming combined heat and power (CHP) systems. Several models of the single-shaft turbines and power electronic devices for them are developed in this chapter. The model of the turbine start is developed as well; the battery is used for this purpose.

Models of plants, utilizing water energy for fabrication of the electrical energy, such as run-of-the-river hydropower plants, ocean wave energy plants, and ocean tide energy plants are developed in Chapter 5. As for the first, models with the permanent magnet synchronous generators with variable rotational velocity and sensorless control are considered, including different optimization methods. The point absorber (PA) and oscillating water column (OWC) are investigated as the devices for transformation of the wave energy into the electrical energy. At this, the PA with various power take-off systems is simulated, including gear rack, magnetic transmission, hydraulic systems, and linear generators. As for the OWC, the models with Wells turbines and impulse turbines are simulated. When ocean tide energy plants are under consideration, both the tidal power plants and tidal current plants are simulated. The conclusion section of the chapter deals with simulation of pump-storage power stations to store large amounts of electrical energy.

Reliable power supply of the autonomous loads is ensured by simultaneous application of RESs of various types together with the sources of electricity that can produce it regardless of environmental conditions. In this way, hybrid systems, or microgrids (MGs), are formed. In Chapter 6, two types of such MGs are simulated: with diesel generators (DGs) and fuel cells (FCs). Since the DG load should not be less than 20–30%, three possible options for

DG operation are proposed: with the ballast load, with the generator disconnected from the diesel using a friction clutch, and with the DG disconnected from the MG, and then switched on under power shortage.

Microgrids consisting of FCs with an electrolyzer, wind generators, photocells, and a battery in various combinations and under different environmental conditions are simulated in this chapter. The integration of individual units both in the DC link and with AC consumer buses is under consideration.

With the employment of SimPowerSystems, a very important problem is the compatibility of the developed models with diverse versions of SimPowerSystems, bearing in mind that every year two versions appear; at this, the new blocks appear, some blocks are declared obsolete, and some of these versions have new features that influence the model performance. It is natural that, bearing in mind the time that is necessary for writing and publishing the book, the author cannot keep up with the latest version. This book is written with the employment of versions R2014a, R2015b, R2016b, and R2019a. In order to run in all these versions, the special model modifications are developed; the respective instructions are given in the book.

This book can be used by engineers and investigators to develop the new electrical systems and investigate the existing ones. It is very useful for students of higher educational institutions during the study of electrical fields, in graduation work and undergraduate's thesis.

MATLAB®/Simulink® is a registered trademark of The MathWorks, Inc. For product information, please contact:

The MathWorks, Inc.
3 Apple Hill Drive
Natick, MA, 01760-2098 USA
Tel: 508-647-7000
Fax: 508-647-7001
E-mail: info@mathworks.com
Web: www.mathworks.com

About the Author

Viktor M. Perelmuter, DSc, earned a diploma in electrical engineering with honors from the National Technical University (Kharkov Polytechnic Institute) in 1958. Dr. Perelmuter earned a candidate degree in technical sciences (PhD) from the Electromechanical Institute Moscow, Soviet Union in 1967, a senior scientific worker diploma (confirmation of the Supreme Promoting Committee by the Union of Soviet Socialist Republics Council of Ministers) in 1981, and a doctorate degree in technical sciences from the Electrical Power Institute Moscow/SU in 1991.

From 1958 to 2000, Dr. Perelmuter worked in the Research Electrotechnical Institute, Kharkov, Ukraine, in the thyristor drive department, where he also served as department chief (1988–2000). He repeatedly took part in putting power electric drives into operation at metallurgical plants. This work was commended with a number of honorary diplomas. Between the years 1993 and 2000, he was the director of the joint venture Elpriv, Kharkov. During 1965–1998, he was the supervisor of the graduation works at the Technical University, Kharkov. Dr. Perelmuter was a chairman of the State Examination Committee in the Ukrainian Correspondence Polytechnical Institute from 1975 to 1985. Simultaneously with his ongoing engineering activity, he led the scientific work in the fields of electrical drives, power electronics, and control systems.

He is the author or a coauthor of 11 books and approximately 75 articles and holds 19 patents in the Soviet Union and Ukraine. Since 2001, Dr. Perelmuter has been working as a scientific advisor in the National Technical University (Kharkov Polytechnic Institute) and in Ltd "Jugelectroproject," Kharkov. He is also a Life Member of the Institute of Electrical and Electronics Engineers.

1

SimPowerSystems Blocks and Units Used for Simulation of the Renewable Energy Systems

1.1 Standard Blocks

The set of SimPowerSystems blocks, Ref. [1], contains the models of rather complex devices and units, whose fields of application are production, transmission, transformation, and utilization of the electric power, electrical drives, and power electronics. Renewable sources of the electrical energy belong to these types of systems, and SimPowerSystems can be successfully used for their creation, simulation, and investigation.

SimPowerSystems operates in the Simulink® environment that allows it to create the system model with the help of the simple procedure "click and drag." After placing the demanded blocks on the model diagram, they are connected according to the system functional diagram. Before the simulation starts, the block parameters must be specified in the block dialog boxes. They can be given as numerical values or letters, the latter is reasonable when the same parameter is present in the several blocks. In this case, the values of parameters are determined either in the option *File/Model Properties/ Callbacks/Init Fcn* or are calculated with a MATLAB® program (MP) with extension .m; this program must be executed before the simulation starts. Its execution can be actuated manually, or the command "run <MPname>" has to be written down in the option mentioned earlier.

This chapter presents a brief description of the SimPowerSystems blocks that are mostly used in the models of renewable energy systems developed in this book. A more complete description of these blocks and others, not mentioned here, is given in Refs. [1–3].

1.1.1 Electrical Sources

A number of models of one- and three-phase electrical sources are included in the library of SimPowerSystems: (1) **DC Voltage Source**, (2) **AC Voltage Source**, (3) **AC Current Source**, (4) **Controlled Voltage Source**, (5) **Controlled Current Source**, (6) **Three-Phase Source** (with the inner impedance),

(7) **Three-Phase Programmable Voltage Source**, and (8) **Battery**. The direct voltage ($\dot{V}$) is assigned for the block of p.1. The blocks of p.2 and p.3 generate the sinusoidal voltage and the current, respectively; the voltage (V) or the current amplitude (A), the phase angle (°), and the frequency (Hz) are specified. The blocks of p.4 and p.5 reproduce the Simulink signal that appears at their inputs.

The block **Three-Phase Source** models a three-phase voltage source that has a Y-connection and R-L internal impedance. The main block parameters are the phase-to-phase root-mean-square (rms) voltage (V), the phase angle (degree), and the frequency (Hz). The block is employed often for simplified modeling of the electric power grid. There are two ways to define R and L. If the option *Specify impedance using short-circuit level* is not checked, the fields appear, in which the values of R (ohm) and L (H) are entered. If this option is chosen, the three-phase short-circuit power (VA) at the base voltage, the value of the base voltage (V), and the ratio X/R are specified. Usually the given rms voltage is taken as the base voltage.

The **Three-Phase Programmable Voltage Source** models a nonstationary and asymmetrical voltage source. The block has several possible configurations of the dialog box. There is an opportunity to vary the main parameters of the output voltage (amplitude, phase, and frequency) during simulation running at the predetermined instants of time. There is also an opportunity to select the type of variation: *Step*, *Ramp* (the linear changing with fixed rate), *Modulation*, and table of *time-amplitude pairs*. Depending on the selected character of variations, the meaning of fields of the dialog box can be changed.

With the use of the item *Fundamental and/or Harmonic generation*, the possibility appears to assign parameters of two harmonics. Four parameters are defined for each harmonic: the harmonic order, its amplitude, its phase, and the type of the sequence (1—for positive, 2—for negative, 0—for zero sequence).

The most complicated model is the model of the **Battery**, which is described in Section 1.2.1.

1.1.2 Loads, Impedances, and Transformers

SimPowerSystems library includes several blocks that contain elements R, L, C connected in parallel or in series: single-phase blocks **Parallel RLC Branch**, **Parallel RLC Load**, **Series RLC Branch**, **Series RLC Load**, and three-phase blocks **Three-Phase Parallel RLC Branch**, **Three-Phase Parallel RLC Load**, **Three-Phase Series RLC Branch**, and **Three-Phase Series RLC Load**. The distinction of branch models from the load models is that for the former, the values of R (ohm), L (H), and C (F) are defined, whereas for the latter, the active, reactive inductive, and reactive capacitive (W or var) powers are given under certain voltage and frequency.

Most of the transformer models use the T-shaped equivalent circuit. The transformer power, the working frequency, the winding voltages, resistances and inductances, as well the magnetization resistance and inductance R_m, L_m, have to be specified. Both SI and pu units can be used for assignment of the resistances and inductances. Saturation is not taken into consideration in the single-phase transformer **Linear Transformer**, whereas it can be made in the models of single-phase **Saturable Transformer** and **Multi-Winding Transformer**. The latter models the transformer with the changing number of windings both on the primary and on the secondary sides. The evenly spaced taps can be added to the first primary or to the first secondary windings. The voltage between the two next taps is equal to the winding total voltage divided by the number of taps plus 1.

The total resistance and the leakage inductance of the winding are evenly spaced along the taps too. This transformer model is usually used for the simulation of the transformer that is controlled when a load changes, with the aim to keep the bus voltage constant. The requested range of the voltage change is ±10–20% of the rated value. The tap number is 6–10. Usually, the transformer has the main primary or secondary winding that is meant for the rated voltage and the adjusting winding with taps that can be connected aiding or opposite to the main winding. With the aiding connection to the primary winding, the output transformer voltage reduces, and with the opposite connection, it increases. When the adjusting winding is connected in series to the secondary one, the effect of switching over is reversed. The examples of utilization of such a transformer are given in Refs. [1–3].

This model gives the possibility to take saturation of the magnetization circuit into account. When it is a case, the magnetization curve is defined as a collection of magnetization current—main flux pairs in the earlier accepted units (pu or SI). The first point is the pair [0 0; ...]. If the residual flux Φ_{res} is modeled, two pairs with zero abscissa are defined: [0 0; 0 Φ_{res}; ...]. A piecewise linear approximation is carried out under simulation.

There is the possibility to model hysteresis. Both major and minor hysteresis loops are simulated using the function *atan*. In order to create the file describing the hysteresis, it is reasonable to use the item *Hysteresis Design Tool* of **Powergui**.

The three-phase transformers **Three-Phase Transformer 12 Terminals**, **Zigzag Phase-Shifting Transformer**, **Three-Phase Transformer (Two Windings)**, and **Three-Phase Transformer (Three Windings)** consist of three single-phase transformers and their main features are the same. The latter two transformers can have the diverse winding connections and can take saturation into consideration: star Y with isolated, grounded, or accessible neutral; delta lagging Y by 30° (D1); and delta leading Y by 30° (D11).

The **Zigzag Phase-Shifting Transformer** model uses three-single-phase three winding transformers; its primary winding is formed by the connection of two windings of the single-phase transformers in the zigzag configuration. In the dialog box, in addition to the usually defined parameters, the phase shift γ of the secondary voltage about the primary voltage is indicated.

Besides the above-mentioned transformers, the blocks of two- and three-winding transformers with inductances of the matrix type model these transformers taking into account the inductive coupling between windings of the different phases. The transformers can have three- or five-limb cores. The modeling of these transformers demands rather detailed information about them that can be received either by computation with known transformer construction or by testing the prototype. Such transformers are not used in the following models.

1.1.3 Transmission Lines, Filters, and Breakers

The transmission line parameters are the active resistance R, the series inductance L, and the parallel capacity C. These parameters are given for a unit of the line length (meters or kilometers). These RLC parameters are evenly distributed along the line length.

The line voltages and the currents change as the electromagnetic waves: the voltage and current values in some point depend not only on the time but also on the position of this point. When the electromagnetic wave reaches the line end, it reflects and propagates in the opposite direction, so that the line voltages and currents are the results of superposition of the incident and reflected waves. However, this effect manifests itself when the line is rather long that is not case in the renewable energy systems, so that the simpler models can be used in these systems.

Instead of the full equations describing the performance of the transmission line, the models with lumped parameters are used. At this, the line model can be considered as a serial connection of several identical RLC sections depicted in Figure 1.1.

The number of sections N depends on the requested maximum frequency of the model, which depends, in turn, on the aim of investigation. For the estimation of the maximum frequency, the following formula is recommended, Ref. [1]

$$f_{\max} = \frac{N}{8d\sqrt{LC}},\tag{1.1}$$

where d is the line length.

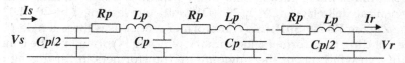

FIGURE 1.1
Transmission line model.

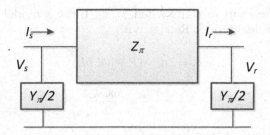

FIGURE 1.2
Equivalent π model of the transmission line.

The f_{max} value of some hundred hertz may be sufficient for line representation when the analysis of the steady state and the transients is carried out, but it may be not sufficient under the investigation of the processes when the line is supplied from the devices of power electronics.

Let $Z = R + j\omega L$ and $Y = j\omega C$, ω is the network angular frequency. The block **PI Section Line** is included in SimPowerSystems that models a single-phase line. The block parameters are the frequency of the transmitted voltage, the line total length, the number of sections, and the values of R, L, and C per 1 km. Each section is modeled with the use of the so-called equivalent π circuit depicted in Figure 1.2. Here (Ref. [4])

$$Z_\pi = Z_c \sinh(\gamma d_s) = Zd_s \frac{\sinh(\gamma d_s)}{\gamma d_s} \tag{1.2}$$

$$\frac{Y_\pi}{2} = \frac{1}{Z_c} \tanh\left(\frac{\gamma d_s}{2}\right) = \frac{Yd_s}{2} \tanh\left(\frac{\gamma d_s}{2}\right) / \left(\frac{\gamma d_s}{2}\right) \tag{1.3}$$

$\gamma = \sqrt{YZ}$, d_s is the section length.
   The value

$$Z_c = \sqrt{\frac{Z}{Y}} \tag{1.4}$$

is the characteristic impedance. Usually, $R \ll \omega L$, then

$$Z_c = \sqrt{\frac{L}{C}}, \tag{1.5}$$

$$\gamma = \alpha + j\beta = \frac{R}{2}\sqrt{\frac{C}{L}} + j\omega\sqrt{LC}. \tag{1.6}$$

Evidently, for the short sections when $\gamma d_s \ll 1$, the π model turns into the nominal π equivalent model, Ref. [4], with

$$Z_\pi = Z d_s = (R + j\omega L) d_s,$$

(1.7)

$$\frac{Y_\pi}{2} = \frac{Y d_s}{2} = \frac{j\omega C d_s}{2}.$$

(1.8)

The block **Three-Phase PI Section Line** models one section of the balanced three-phase line by using the equivalent π model. The specified parameters are the positive sequence impedances and zero sequence impedances per 1 km: r_1, r_0, l_1, l_0, c_1, and c_0. After multiplication by the line length d_s, the full impedance values are obtained: R_1, R_0, L_1, L_0, C_1, and C_0. By simulation, these values of sequences are transformed into the self and mutual parameters of the model that consists of the elements described earlier. The line is supposed to be balanced (all the diagonal elements are the same, and all the nondiagonal elements are the same); in such a case, the impedance matrix is diagonal and has only two different elements that are called positive sequence impedance and zero sequence impedance. This block is described in more detail in Refs. [1, 3]. The impedance values are determined by the line construction; the **Powergui** option *RLC Line Parameters* can be used for this purpose, Refs. [1, 2].

Together with the overhead transmission line, SimPowerSystems gives the possibility to calculate the parameters of the cable line with the command *power_cableparam*, which computes the inductances and capacitances of the system, consisting of N-shielded radial cables. It is supposed that the cable consists of the copper conductor with an outer screen and utilizes the polyethylene insulation.

Since the cable lines are usually not very long, the π model depicted in Figure 1.3 can be used for cable modeling. Details of the cable parameter computation and examples of utilization are given in Ref. [3].

The other blocks can be used for transmission line simulation as well. For example, the block **Three-Phase Mutual Inductance Z1–Z0** is employed in the demonstration model *power_wind_ig* for creating the grid with three-phase short-circuit power of 2.5 GVA at the voltage of 120 kV. This block models a three-phase balanced inductive and resistive impedance with mutual coupling between phases. The block parameters are the active and reactive self and mutual impedances of the positive and zero sequences. These parameters are calculated usually during computation of the three-phase line parameters. In the above-mentioned model $R_0/R_1 = L_0/L_1 = 3$ is accepted.

For the short line, the simple R-L circuits can be used.

The block **Three-Phase Harmonic Filter** is intended for simulation of the parallel filters in the power systems that are used for reduction of the voltage distortions and for the power factor correction. Usually, the band-pass

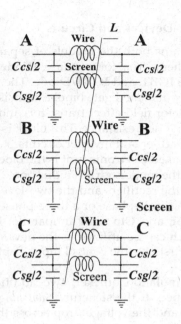

FIGURE 1.3
The π cable model.

filters are used that are tuned for low-order harmonic filtering: the 5th, the 7th, the 11th, and so on. The band-pass filters can be tuned for filtering one frequency (single-tuned filter) or two frequencies (double-tuned filter). High-pass filters are used for filtering a wide range of frequencies. A special type of high-pass filter, the so-called C-type high-pass filter, has some merits over the other types of high-pass filters. The filters contain *R*, *L*, and *C* elements. The values of these elements are determined by the following data: the filter type, the reactive power at the rated voltage, the tuned frequency or frequencies, and the quality factor. The examples of filter parameter computations and obtained frequency characteristics are given in Ref. [2].

The blocks **Breaker** and **Three-Phase Breaker** model the switches that can switch over by the inner timer or by the external signal Simulink (0—Off, 1—On). At this, the actual contact opening takes place after the conducted current crosses zero level. So, they cannot work in the direct current circuits; the block **Ideal Switch** can be used in such cases that, unlike the **Breaker**, permits a commutation with current.

The **Three-Phase Breaker** consists of the three single-phase ones; only part of the contacts can switch over; the rest of contacts remain in the initial state. The blocks contain *R-C* snubber that must be used when these blocks are set in series with the block which are modeled as a current source, for examples, an inductor.

1.1.4 Power Electronic Devices and Circuits

SimPowerSystems contains both the models of separate electronic devices and the units using them. The formers are **Diode, Thyristor, Detailed Thyristor, Gto, Mosfet, IGBT**, and **IGBT/Diode**. The latter block is the simplified model of the pair insulated gate bipolar transistor (IGBT) (or GTO, or metal-oxide-semiconductor field effect transistor) and antiparallel **Diode**; it has only one parameter: inner resistance. This block is commonly used when modeling the various rectifier and inverter circuits.

The most intensively used electronic unit is the block **Universal Bridge**. It is present in almost all the models given in this book. This block is the basis for building models of the rectifiers and the two-level inverters. It can have one arm or can model a single-phase or a three-phase bridge converter. The blocks **Diode, Thyristor**, and **Gto** with antiparallel **Diode, Mosfet, IGBT/ Diode**, and **Ideal Switch** can be utilized as the devices in this bridge. In all these cases, except the first one, the control input g appears for receiving the vector of gate pulses.

In order to make simulation quicker, two additional modifications of this block were developed. If the structure *Switching-function based VSC* is selected, the transients and the voltage drop across the inverter devices are neglected, the output voltage space vector components are proportional to the voltage U_d across the capacitor in the inverter direct current (DC) link; at this, proportionality factors depend on the gate pulse combination. In fact, the block contains two controlled voltage sources that fabricate the space vector and the controlled current source that forms the DC link current. The gate pulses come to this block as before, and the harmonic contents of the output voltage remain.

When the second structure *Average-model-based VSC* is chosen, the output voltage is proportional to the input signal U_{ref}; moreover, the proportionality factor depends on the capacitor voltage. At this, the simulation runs faster, but the harmonic contents are not preserved. This possibility can be used for the simulation of the system that contains a lot of voltage source converters (VSCs).

The block **Three-Level Bridge** is utilized in the models of three-level neutral point clamped (NPC) power converters. **MOSFET/Diodes, GTO/Diodes, IGBT/Diodes**, and **Ideal Switches** can be used in this block.

Beginning from the version 2015b, the Power Electronics library contains seven new blocks that represent common power converter topologies. These blocks and the functions, which they are intended for, are shown in Figure 1.4. The diagrams of the boost, buck, and two-direction boost–buck converters are depicted in Figure 1.5. One can see that their electronic circuits really correspond to the circuits of the blocks with the same names that are shown in Figure 1.4. Of course, these circuits can be implemented without these new blocks, how it is made, for example, in the some models in Ref. [3]; but the new blocks have merits because they can operate in three

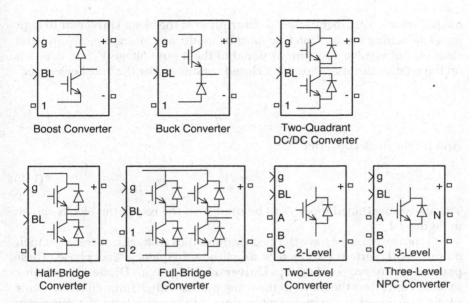

Boost Converter   **Buck Converter**   **Two-Quadrant DC/DC Converter**

Half-Bridge Converter   **Full-Bridge Converter**   **Two-Level Converter**   **Three-Level NPC Converter**

FIGURE 1.4
Pictures of the additional blocks of power electronics.

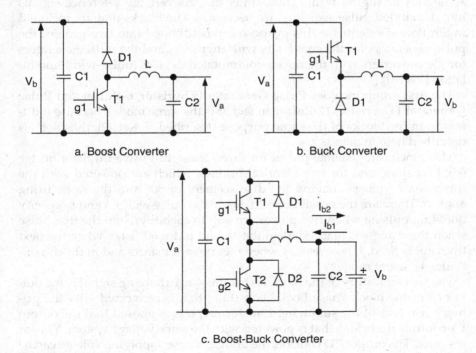

a. Boost Converter

b. Buck Converter

c. Boost-Buck Converter

FIGURE 1.5
Diagrams of DC/DC converters.

modes, which were mentioned in description of the block **Universal Bridge**: usual switching mode, *Switching-function* mode, and *Average-model*. In latter case, output voltage V_b is proportional to the input voltage V_a and depends on the relative duration d of IGBT closed condition. For the boost converter

$$V_b = \frac{V_a}{1-d}. \tag{1.9}$$

And for the buck converter

$$V_b = dV_a. \tag{1.10}$$

The above-mentioned modes can be used with the rest of the blocks shown in Figure 1.4.

The converters must have the electric quantities (voltages, currents) at their outputs with certain parameters: amplitude, frequency, and phase. These parameters are provided (except **Universal Bridge** with **Diodes**) by sending to these converters the trains of the firing pulses with definite characteristics: frequency, duration, arrangement in time. At the same time, the references for the requested electric parameters are generated by the control system in analogous or digital form. Thus, units that convert the reference signals into demanded pulse sequence are necessary. The blocks that are included in SimPowerSystems for this purpose can be divided into two groups: the pulse generators for the converters with thyristors and the pulse generators for the converters with the forced-commutated devices (pulse-width modulation [PWM]).

The first group includes **Pulse Generator (Thyristor, 6-Pulse)** and **Pulse Generator (Thyristor, 12-Pulse)**, in fact, it is the same model. Because it differs from the blocks of the same purpose described in Ref. [2], this block is described here in detail.

This block can generate pulses for three-phase full-wave thyristor bridge (six thyristors) and for two identical bridges, which are powered with the three-phase voltages having the displacement of 30° and the same firing angle α. There are the possibilities to define the pulse width Δ and to specify doubling pulsing when two pulses are sent to each thyristor: the first pulse when the α angle is reached, then the second pulse 60° later, when the next thyristor is fired. It is necessary when the converter starts and in the discontinuous current mode.

The special feature of the block is that the synchronizing signal is the output ωt of the block Phase Lock Loop (PLL) that is connected with the primary winding of the supplying transformer; it is supposed that the output **PY** controls the bridge that is powered with the same voltage system, Y/Y for instance. The output **PD** controls the bridge, whose supplying voltage can be Y/D1 or Y/D11. The block structure is shown in Figure 1.6, and the fabrication of pulses is shown in Figure 1.7.

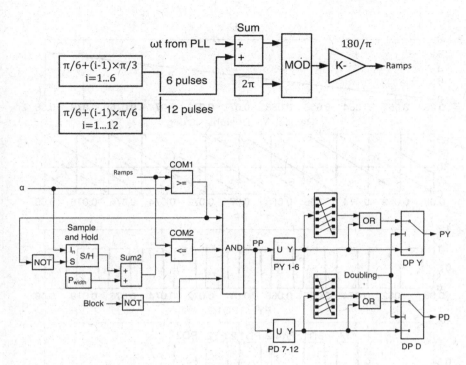

FIGURE 1.6
Diagram of the **Pulse Generator**. (a) Fabrication of the *Ramp* signals; (b) pulse generation.

The ωt signal is the angle varying between 0 and 2π radians, synchronized with zero crossings of the phase primary voltage U_a of the converter transformer. Because the natural T_1 firing (firing in the diode bridge) takes place, when the phase voltages $U_a = U_c$, the internal ωt signal for T_1 crosses zero 30° later. Signals for the other thyristors are shifted relatively one to another by 60°, the internal ωt signals for the second bridge are shifted relatively ones for the first bridge by 30° in the direction that depends on the selected *Delta winding connection*. Fabrication of these signals designated as *Ramps* is shown in Figures 1.6a and 1.7b (these signals vary within 0–360°, only part of these signals 0–90° is shown in Figure 1.7b).

Further pulse generation is depicted in Figure 1.6b. *Ramps* are compared with the angle α by comparing blocks **COM1**. When appropriate signal *Ramp* is getting more than α, the logic 1 appears at the output of **COM1**; at this, until *Ramp* < $\alpha + \Delta$, the logic signal 1 will be kept at the output of **COM2**, and, with the stipulation that signal *Block* = 0, the pulse with width of Δ (designated as PP) appears at the output of the block AND. It will be 12 pulses during the period of the supplying voltage for both converter bridges (see Figure 1.7c). With the help of the diverse **Selectors**, logic gates, and **Switches**, the incoming pulses are directed to the appropriate thyristors. The pulses

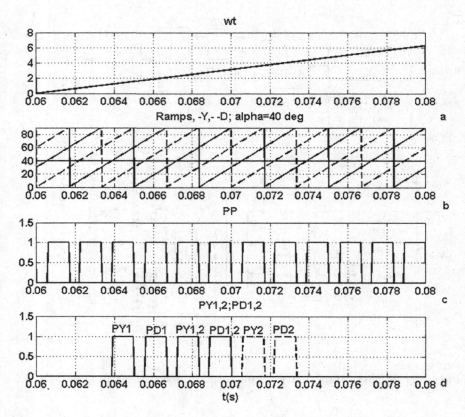

FIGURE 1.7
Pulse generation for two full-wave thyristor bridges.

for the thyristors T_1 and T_2 of both bridges are shown in Figure 1.7d. One can see that when the pulse comes to T_2, the repeated pulse appears at T_1 as well (doubling pulse is *On*). The pulses for the second bridge are shifted about the pulses for the first one by 30°.

In the previous versions of such types of the blocks, synchronization was carried out by the supplying voltages directly; explicit definition of the voltage frequency had to be set in the dialog box. It can result in inaccurate reproduction of the desired angle α in the weak grids, where the frequency can vary appreciably. There were variants with the frequency assignment by the external signal from some measuring devices, but it makes the units more complicated. In the block under consideration, explicit definition of the voltage frequency is not necessary, but PLL has some response time that, under abrupt change of the supplying voltages, can result in erroneous synchronization; it is especially dangerous in the inverter mode.

The second group includes **PWM Generator (2-Level)**, **PWM Generator (3-Level)**, and **SVPWM Generator (2-Level)**. These blocks do not differ essentially from the blocks analogous by functions and described in Refs. [2, 3]; they have only slight circuit distinctions. Therefore, they are not described here in detail. Remember that the first two blocks use a comparison of the three-phase reference (or modulating) signal U_r with an isosceles triangle carrier waveform U_t and the fabrication of the pulses when U_r crosses U_t. Signal U_t changes within $(-1, 1)$; therefore, signal U_r has to be in the same range, in order to stay in the line region. At this, the relationship between DC voltage U_d and the effective value of the phase-to-phase voltage of the fundamental harmonic U_{l1} is

$$U_{l1} = \frac{\sqrt{6}}{4} m U_d = 0.612 m U_d, \qquad (1.11)$$

where m is modulation index, $0 \le m \le 1$. The linear range can be extended by addition of the zero sequence triple harmonic of the appropriate amplitude to the three-phase reference signal; it means that the same signal is added to all three phases. This zero-sequence signal does not appear in the phase-to-phase voltages. The SimPowerSystems library includes the block **Overmodulation** that carries out this operation. The block implements three options. In the *Third harmonic* option, the third-harmonic signal V_0 subtracted from the original signal is calculated as

$$V_0 = \frac{|U|}{6} \sin\left[3(\omega + \angle U)\right], \qquad (1.12)$$

where U is the reference signal.

In the *Flat Top* overmodulation option, the portion of the three-phase input signal exceeding values ± 1 is computed. Three resulting signals are then summed up and removed from the original signal U_r.

In the Min–Max overmodulation option, the additional signal is

$$\Delta V = -0.5 \times \left[\max\left(V_{rabc}\right) + \min\left(V_{rabc}\right)\right]. \qquad (1.13)$$

The block outputs a value between -1 and 1 in all these cases; it is instead of Equation (1.11)

$$U_{l1} = \frac{1}{\sqrt{2}} m U_d = 0.707 m U_d. \qquad (1.14)$$

The outputs of this block are displayed in Figure 1.8 when the amplitude $|U_r| = 1.15$.

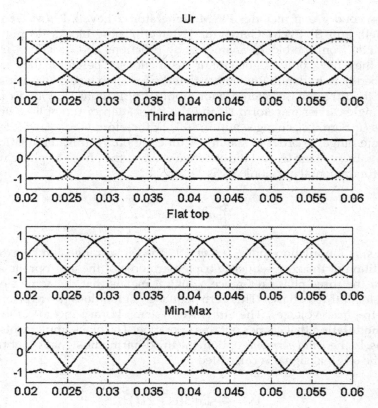

FIGURE 1.8
Outputs of the block **Overmodulation** in different modes.

The block **SVPWM Generator (2-Level)** uses the method of the space vector modulation (SVM), which is a digital method in principle. A system operates with the time period T_c; several VSI states are selected during T_c. For SVM, the control system forms the requested output voltage vector u—in the Cartesian or in the polar coordinates. Two adjacent VSI states and one of zero states are used in order to receive the vector u; the formulas for calculation of the time intervals, during which these states are active, are given in Refs. [2, 3].

The block **PWM Generator (DC-DC)** is intended for the control of the buck or boost DC converter. The input signal is compared with the sawtooth waveform that changes within (0–1); the value of the input D determines the percentage of the pulse period when the output is on. The buck and boost type converters are used widespread in the renewable electric sources. When PWM is developed by the user, who cannot employ the standard PWM blocks described earlier, the sources of the sawtooth or the triangle waveforms are often needed. Such sources are included in the SimPowerSystems library: **Triangle Generator** and **Sawtooth Generator**.

1.1.5 Electric Generators

Three-phase induction generators (IGs), both with squirrel cage IG (SCIG) and with wound rotor IG (WRIG or doubly fed induction generator [DFIG]), as well as with wound rotor synchronous generator (WRSG) or with permanent magnet rotor synchronous generator (PMSG), are widely applicable in the RESs. The models of these generators in SimPowerSystems have some identical features. Modeling is carried out with the transformation of the three-phase generator to two-phase one; after computation, the machine currents are transformed back in the three-phase system and, with the help of the **Controlled Current Sources**, are made for the external circuits. The models using either International System of Units (SI units) or pu units can be chosen in SimPowerSystems, except of the PMSG, but the simulation is carried out always in pu units. For the PMSG, the electromechanical equations are used in SI units as

$$J\frac{d\omega_m}{dt} = T_e - f\omega_m - T_m, \tag{1.15}$$

$$\frac{d\theta_r}{dt} = Z_p\omega_m. \tag{1.16}$$

where
J is a total moment of inertia of the PMSG and connected parts
f is a viscous friction factor
T_e is the PMSG electromagnetic torque
T_m is a load torque, $T_m < 0$ in generator mode
θ_r is the rotor electrical angle, *rad*
Z_p is the number of pole pairs.

It should be noted that the value of the mechanical angle appears at the Simulink output of the block m that is less than the electric angle in Z_p times.
    For the other models, the electromechanical equations are used in pu as

$$2H\frac{d\omega_r}{dt} = T_e - f\omega_r - T_m, \tag{1.17}$$

where T_m is given in pu as well; the inertia constant

$$H = \frac{J\omega_b^2}{2Z_p^2 P_b} \tag{1.18}$$

where P_b is the nominal generator power (VA), the speed is defined in the ratio of the base speed ω_b that is usually the stator nominal frequency ω_s, $\omega_r = \omega_m/\omega_b$. The rotor electrical angle is defined as

$$\frac{d\theta_r}{dt} = \omega_b\omega_r \tag{1.19}$$

The T-shape equivalent circuits are used to model an IG. It means that the voltage of the DFIG rotor winding is equal to the stator winding under $\omega_m = 0$. Usually it is not the case, therefore, when it is necessary to observe the real voltages and currents in the rotor circuits, the virtual ideal transformer must be used connected with the rotor winding; the turn ratio of this transformer must be equal to the DFIG winding ratio.

There are three options for choice of the reference frame to convert input voltages to the *dq* reference frame: if ω is the rotation speed of the selected reference frame, then it can be $\omega = 0$ (stationary reference frame), $\omega = \omega_s$, ω_s is a synchronous speed, 1 in pu (synchronous reference frame), and $\omega = \omega_r$ (reference frame connected with the rotor).

Let's introduce the vector $\mathbf{\Psi} = [\Psi_{qs}, \Psi_{ds}, \Psi_{qr}, \Psi_{dr}]$, and the vectors $\mathbf{U}$ and $\mathbf{I}$ that are arranged by the same way, the diagonal 4×4 matrix of the resistances $\mathbf{R}$, the matrix of the inductances $\mathbf{L}$

$$\mathbf{R} = \begin{bmatrix} R_s & 0 & 0 & 0 \\ 0 & R_s & 0 & 0 \\ 0 & 0 & R_r & 0 \\ 0 & 0 & 0 & R_r \end{bmatrix} \quad \mathbf{L} = \begin{bmatrix} L_m + L_{ls} & 0 & L_m & 0 \\ 0 & L_m + L_{ls} & 0 & L_m \\ L_m & 0 & L_m + L_{lr} & 0 \\ 0 & L_m & 0 & L_m + L_{lr} \end{bmatrix} \tag{1.20}$$

and also the matrix

$$\mathbf{W} = \begin{bmatrix} 0 & \omega & 0 & 0 \\ -\omega & 0 & 0 & 0 \\ 0 & 0 & 0 & \omega - \omega_r \\ 0 & 0 & \omega_r - \omega & 0 \end{bmatrix} \tag{1.21}$$

where R_s and R_r are the active resistances of the stator and rotor windings, respectively; L_{ls} is the stator leakage inductance, L_m is the magnetizing inductance, L_{lr} is the rotor leakage inductance, and ω is rotation speed of the reference frame. Then the equation for $\mathbf{\Psi}$ can be written as (Ref. [3])

$$\frac{d\mathbf{\Psi}}{dt} = (\mathbf{U} - \mathbf{R}\mathbf{L}^{-1}\mathbf{\Psi} - \mathbf{W}\mathbf{\Psi})\omega_b. \tag{1.22}$$

After computation of this equation, the currents are found as

$$\mathbf{I} = \mathbf{L}^{-1}\mathbf{\Psi}. \tag{1.23}$$

Equations (1.22) and (1.23) are computed in the model of an IG. The electromagnetic torque is calculated as

$$T_e = \Psi_{ds} i_{qs} - \Psi_{qs} i_{ds}. \tag{1.24}$$

The following parameters are specified in the dialog box of the IG: rotor type, reference frame, nominal power, stator voltage, and its frequency in addition to the above-mentioned parameters: L_{ls}, L_m, L_{lr}, R_s, and R_r—in SI or in pu— J or H, f, and Z_p.

Saturation can be taken into consideration. The magnetization curve is modeled for the total magnetic flux in the no-load machine. It is carried out in the following way. The fluxes Ψ_{md} and Ψ_{mq} may be defined as

$$\Psi_{mq} = L_a \left(\frac{\Psi_{qs}}{L_{ls}} + \frac{\Psi_{qr}}{L_{lr}} \right) \tag{1.25}$$

$$\Psi_{md} = L_a \left(\frac{\Psi_{ds}}{L_{ls}} + \frac{\Psi_{dr}}{L_{lr}} \right) \tag{1.26}$$

$$L_a = \left(\frac{1}{L_m} + \frac{1}{L_{ls}} + \frac{1}{L_{lr}} \right)^{-1}. \tag{1.27}$$

Then the saturated inductance L_m is calculated as a function f of the flux module

$$I = f\left(\sqrt{\Psi_{md}^2 + \Psi_{mq}^2} \right), \qquad L_m = \frac{\sqrt{\Psi_{md}^2 + \Psi_{mq}^2}}{I}; \tag{1.28}$$

the function f is defined by the $2 \times n$ matrix in the dialog box, where the first row defines the stator currents and the second row defines the stator voltages under these currents in the no-load condition; n is the number of points of the magnetization curve. The first point is the point where the effect of saturation begins. This value L_m is used afterward in formulas (1.20), (1.22), (1.23), and (1.27).

Pay attention to the next special feature. In the wind generator (WG) and some other RESs, IGs with squirrel cage rotor are often employed. In order to such a generator can operate in an isolated network and in some other cases, it has to self-excite. The library IG model does not provide a self-excitation function. There is the possibility to define the initial stator currents that, when IG starting, lead to self-excitation, but the obtaining process disagrees with the actual process, when self-excitation is provided by the residual flux. Two opportunities to provide self-excitation are considered in Ref. [3]. The first one is to develop a special IG model that takes into account the self-excitation process. For that, it is sufficient to have the possibilities to set the initial stator and rotor fluxes as the residual ones.

The second opportunity is to use the library model of IG but with wound rotor. Its parameters are the same as for the first case. For self-excitation modeling, two voltage-controlled sources, placed in the subsystem **SelfExcit**, are connected to the rotor windings. The voltage amplitude is about 2%; the frequency is proportional to the IG speed. When the self-excitation process comes to the end, the rotor windings are shorted, and subsequently the IG operates as squirrel cage one.

The model of the SG is considered further. It is supposed that SG with salient poles, except of the field winding, has two damper windings on the rotor, whose axes are aligned along the direct and quadrature axes. It is supposed that round rotor SG has two damper windings in the latter axis. Only the equations for salient pole SM are given further.

The SG modeling is carried out in the rotating reference frame aligned with the rotor. The parameters of electrical circuits of the SG model are the stator winding resistance and leakage inductance R_s and L_{sl}, with i_{qs} and i_{ds} as its currents, Ψ_{qs} and Ψ_{ds} as its flux linkage; the quadrature damper winding resistance and leakage R_{kq} and L_{lkq}, with i_{kq} as its current, and Ψ_{kq} as its flux linkage; the direct damper winding with parameters R_{kd} and L_{lkd}, with i_{kd} as its current, and Ψ_{kd} as its flux linkage; the excitation winding with parameters R_f and L_{lf}, and Ψ_f and i_f as its flux linkage and current. Let L_{mq} and L_{md} are the q-axis and d-axis magnetizing (main) inductances.

If to designate $\mathbf{V} = [U_{qs}\ U_{ds}\ 0\ U_f\ 0]$ where the first two components are the stator voltage components, U_f is the excitation voltage, $\mathbf{\Psi} = [\Psi_{qs}\ \Psi_{ds}\ \Psi_{kq}\ \Psi_f\ \Psi_{kd}]$, $\mathbf{I} = [i_{qs}\ i_{ds}\ i_{kq}\ i_f\ i_{kd}]$, $\mathbf{R}$ is the diagonal matrix of the resistances; and to designate as $\mathbf{W}$ the matrix that consists of zero elements except $w(1, 2) = \omega_r$ and $w(2, 1) = -\omega_r$, the equations for the flux linkages can be written as

$$\mathbf{V} = \mathbf{R}\mathbf{I} + \frac{d\mathbf{\Psi}}{dt} + \mathbf{W}\mathbf{\Psi}. \tag{1.29}$$

The equations for the flux linkages are

$$\Psi_{qs} = L_q i_{qs} + L_{mq} i_{kq}, \ L_q = L_{sl} + L_{mq}$$

$$\Psi_{ds} = L_d i_{ds} + L_{md} i_{kd} + L_{md} i_f, \ L_d = L_{sl} + L_{md}$$

$$\Psi_{kq} = L_{kq} i_{kq} + L_{mq} i_{qs}, \ L_{kq} = L_{lkq} + L_{mq} \tag{1.30}$$

$$\Psi_f = L_{fd} i_f + L_{md} i_{kd} + L_{md} i_{ds}, \ L_{fd} = L_{lf} + L_{md}$$

$$\Psi_{kd} = L_{md} i_f + L_{kd} i_{kd} + L_{md} i_{ds}, \ L_{kd} = L_{lkd} + L_{md}$$

which can be written as

$$\mathbf{\Psi} = \mathbf{L}\mathbf{I}. \tag{1.31}$$

After substitution in Equation (1.29), we get Equation (1.22); the currents are computed as in Equation (1.23). The torque is calculated by Equation (1.24).

Of course, in addition to the above-mentioned parameters, rotor type, nominal power, stator voltage, and its frequency, J or H, f, and Z_p, must be given in the dialog box of the SG. It should be borne in mind that for the SG, as distinct from the IG, where the torque $T_m < 0$ in the generator mode, the power P_m must be positive in this mode.

A more detailed description of the SG model is given in Refs. [1–3]. It is worth noting that the model of **Simplified Synchronous Machine** is present in SimPowerSystems (Refs. [1–3]). This model is reasonable to employ in the models of the wind parks, when aggregating does not carry out or fulfills not all-round.

The PMSG has only a stator winding; an excitation is carried out with the permanent magnets mounted on the rotor. One distinguishes the surface magnet machines (SPMSG) and the interior magnet machines (IPMSG). The former have magnets that are glued on the rotor surface, whereas the latter have magnets that are mounted inside the rotor. Since the relative permeability of PM material is very close to 1, the presence of PM does not have influence on the stator winding inductance. Therefore, $L_d = L_q = L_s$ for SPMSG, whereas $L_d < L_q$ for the IPMSG, because of the bigger air gap in the direction of axis d, owing to the necessity to place magnets. PMSG is described by four electromagnetic parameters: the stator winding resistance R_s, the inductances L_d and L_q, the rotor flux Ψ_r, and also the pole pair number Z_p.

The PMSG is modeled in the reference frame that rotates in synchronism with the rotor. The PMSG equations may be obtained from the equations of the usual SG as (Ref. [1–3]):

$$v_{ds} = R_s i_{ds} + L_d \frac{di_{ds}}{dt} - \omega_s L_q i_{qs} \tag{1.32}$$

$$v_{qs} = R_s i_{qs} + L_q \frac{di_{qs}}{dt} + \omega_s \left(L_d i_{ds} + \Psi_r \right). \tag{1.33}$$

PMSM torque may be written as

$$T_e = 1.5 Z_p \left[\Psi_r i_{qs} + \left(L_d - L_q \right) i_{ds} i_{qs} \right]. \tag{1.34}$$

Because for SPMSG the torque is independent of i_{ds}, it is preferably to have $i_{ds} = 0$, in order to decrease the SPMSG heating and to increase the load-carrying capacity. It follows from Equation (1.34) that, if to take $i_{ds} < 0$ for IPMSG, its torque will be bigger, because of $(L_d - L_q) < 0$; but at this, maximal value i_{qs} decreases, because it must be

$$i_s = \sqrt{i_{ds}^2 + i_{qs}^2} \le i_0, \tag{1.35}$$

where i_0 is the maximal permissible current of the PMSG or the inverter. Therefore, the optimal value of i_{ds} exists. It is convenient to introduce the relative quantities (Refs. [2, 3]): $I_q = L_d i_{qs}/\Psi_r, I_d = L_d i_{ds}/\Psi_r, I_0 = L_d i_0/\Psi_r$, $dl = \left(L_q - L_d \right)/L_d, m = T_e L_d/1.5 Z_p \Psi_r^2$, then it follows from Equation (1.34)

$$m = I_q(1 - I_d \, dl). \tag{1.36}$$

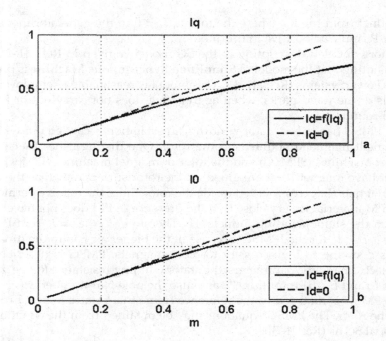

FIGURE 1.9
Dependencies I_q and I_0 on m for IPMSG.

The quantities I_q and I_d are chosen so that the torque would be maximal with the given current I_0. After substitution of $I_q = \sqrt{I_0^2 - I_d^2}$ in Equation (1.36), maximization by I_d, and substitution in the final expression $I_0 = \sqrt{I_q^2 + I_d^2}$, the optimal relationship will be received as

$$I_q^2 = I_d^2 - \frac{I_d}{dl}. \qquad (1.37)$$

Dependences of the currents I_q (Figure 1.9a) and I_0 (Figure 1.9b) on m are shown in Figure 1.9 with $dl = 1$. One can see that, with utilization of the dependence of I_d on I_q according to Equation (1.37), the total current decreases.

This possibility is implemented in the model **Wind_PMSG_4N**.

1.1.6 Powergui

The graphical user interface **Powergui** gives a number of opportunities to choose the required simulation conditions and to gain more information about the simulation results. This block is used for saving model data, so that its presence on the model diagram is obligatory. Only one such block may be placed on this diagram; its name cannot be changed. But when there are some models of the electrical circuits on the same diagram, not connected one with other, each model can have a separate **Powergui**. The functions and the dialog box of this block change in the different versions of SimPowerSystems.

The block has three item pages. The first one *Solver* gives the opportunity to select the method of simulation (continuous, discrete, or Phasor). If the first method is chosen, the additional possibilities appear to make the simulation faster by omitting some details of switching devices. When the discrete method of simulation or the Phasor method is selected, the sampling time and the discretization method or the frequency must be specified, respectively.

The opportunity appeared in the latest versions to choose the option *Interpolate* in the mode *Discrete/Tustin*. When the solver finds out that the gate pulse transitions of power electronic devices occur between two sample times, it corrects computations using interpolation; this, at definite conditions, gives the possibility to increase the sampling time and simulation speed. The advantages of this option are shown in the simple demonstration model **power_buck**. But it is necessary to have in mind that, with increasing model complexity, availability of several power electronic devices, controllers, etc. the merits of this option decrease essentially.

The view of the second page of this block for R2019a is displayed in Figure 1.10. The item *Steady-State* gives the possibility to find the steady-state

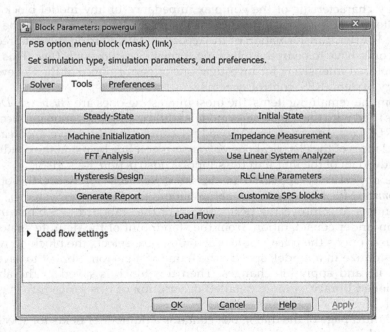

FIGURE 1.10
The **Powergui** second page.

values of the variables in the vectors **x**, **u**, and **y**, which form the linear part of the model, steady-state voltages, and currents of measurement blocks and sources, as well as steady-state voltages and currents of nonlinear blocks of the model.

If the model contains power electronic switches (diodes, thyristors, IGBTs, and so on) the steady-state values are computed considering them as open circuit.

The item *Initial States* make it possible to display the initial values of the electrical state variables (inductor currents and capacitor voltages) of a model, as well as to modify the initial electrical states in order to start the simulation from any initial conditions.

The item *Load Flow* gives the opportunity to run the simulation, beginning from the steady state. Under realization of this item, the system of nonlinear equations, linking voltages, currents, and powers in all nodes of the electrical system in the steady state, is solved by an approximate method; subsequently, the computed values of voltages, currents, and powers are fed into the model automatically, and the simulation starts. A more detailed description of this procedure is given in Refs. [1–3], where the examples of employment of this and the next item are given.

The neighbor item *Machine Initialization* is the simplified version of the previous one.

The item *Impedance Measurement* gives the opportunity to build the frequency characteristic of the complex impedance of any model block. The next item *FFT Analysis* is intended for the frequency analysis with the help of the Fourier transformation of the recorded data. During the data record, the scopes have to be in the following format: *Structure with Time*. This item is employed intensively for investigations of performance of the Renewable Energy System (RES) control systems.

From the remaining items, the most interesting ones are *Hysteresis Design* that is used with the transformer models while taking saturation into account and *RLC Line Parameters* that is used for computing the active and reactive impedances of the electrical power transmission line by using the conductor data, the data of the line, and the support construction, Refs. [1–3].

The opportunity appeared in the latest versions to choose the option *Customize SPS blocks* (SPS is abbreviation for SimPowerSystems). This option lets to improve the performance of the standard SPS block in the system under consideration, from the standpoint of the definite customer. The user opens the *power_customize* dialog box, selects the block he wants to customize in a model, specify the name of a custom library to save the block in, and apply his changes. Then this block is saved in the above-mentioned library. A more detailed description of this procedure is given in Ref. [1].

The third page of the dialog box contains a number of fields for specification of the parameters that can be used under the computation of the load flow for the model: the frequency, model tolerance under the solution of the nonlinear equations with the approximating method, and so on.

1.1.7 Control and Measuring Blocks

There are a number of blocks in the folder *Control and Measurements*, which make building of the control system lighter. The short description of the blocks that are used in RES is given next.

Block **PLL** has been mentioned already under the description of **Pulse Generators**. Its diagram is depicted in Figure 1.11. The q component of the voltage space vector V when the reference frame rotates with the angle frequency ω may be written as (Ref. [5])

$$V_q = \frac{2}{3}\left[V_a \cos x + V_b \cos\left(x - \frac{2\pi}{3} \right) + V_c \cos\left(x + \frac{2\pi}{3} \right) \right], \qquad (1.38)$$

$x = \omega t$. Substituting $V_a = V_m \sin \theta_1$, $V_b = V_m \sin (\theta_1 - 2\pi/3)$, $V_c = V_m \sin (\theta_1 + 2\pi/3)$ with θ_1 as the actual position of the voltage space vector and $x = \theta_2$, where the last quantity is the estimation of θ_1, it follows $V_q = V_m \sin (\theta_1 - \theta_2)$; this value is equal to zero when the estimation is true ($\theta_2 = \theta_1$) and can be used for angle tuning. The block input quantities are the phase voltages V_a, V_b, and V_c; the outputs are the estimations of the frequency $f = \omega_2/2\pi$ and θ_2, $\omega_2 = d\theta_2/dt$.

The block **First-Order Filter** realizes the transfer function of the first-order, whose form depends on the selected filter type: *Low-pass* or *High-pass*. The time constant of the filter T is specified in the dialog box.

The block **Second-Order Filter** gives more possibilities for processing the signals. The filter types *Low-pass*, *High-pass*, *Band-pass*, and *Band-stop* may be selected. The filter natural frequency f_n and the damping factor ξ are specified in the dialog box. For *Low-pass* and *High-pass* filters, the amplitude–frequency characteristics are equal to $1/(2\xi)$ at f_n; this characteristic has a maximum at this frequency for *Band-pass* and a minimum for *Band-stop* filters.

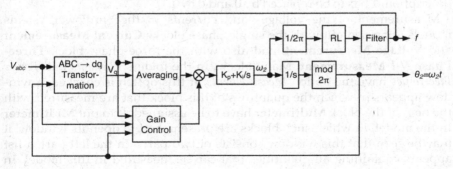

FIGURE 1.11
PLL block diagram.

Six blocks for signal transformation are placed in this folder, namely: Clarke transformation $(a-b-c) \rightarrow (\alpha-\beta-0)$ and vice versa; Clarke–Park transformation with angular position of the rotating frame given by input ωt $(\alpha-\beta-0) \rightarrow (d-q-0)$ and vice versa; Park transformation also with angular position of the rotating frame given by input ωt $(a-b-c) \rightarrow (d-q-0)$ and vice versa. The rotating frame may be aligned with the phase A axis or may be 90° behind.

A number of blocks are used in the discrete systems for forming different signals, storing of the fabricated quantities, signal distribution over time, etc. They are as follows: the block **Bistable** is an R–S flip-flop that sets the output $Q = 1$ under $S = 1$ and $R = 0$, resets $Q = 0$ under $S = 0$ and $R = 1$, and holds Q under $S = 0$ and $R = 0$. The state when $S = 1$ and $R = 1$ can be assigned.

The output of the block **Sample and Hold** follows its input until the logic signal 1 holds at the block input S $(S = 1)$. When S is going to 0, the block output holds its input at this instant of time. The initial output value can be specified in the dialog box.

For the block **Monostable**, the direction of the input change is specified, for which the blocks respond (Edge: *Rising*, *Falling*, or *Either*). If this event takes place, the block outputs the logical 1 during the time that is fixed as *Pulse duration*.

The block **Edge Detector** continuously compares the current value of the input signal with the previous one. The block can be tuned for the generation of output logical signal 1 by increasing or decreasing the input signal, or when there is any inequality between the current and previous values. When the specified event takes place, the logical input signal 1 appears, whose duration is equal to one integration step.

The block **On/Off Delay** is intended for the delay of the input logical signal for the specified time *Time Delay*. For *On delay*, the input = 1 appears at the output after *Time Delay* and is held in this state. When the input is going to 0, the output is going to 0 without delay. For the *Off delay* mode, in this description, 1 has to be replaced by 0 and 0 by 1.

Measurement of the voltages and currents in the SimPowerSystems models is carried out with the single-phase blocks **Current Measurement** and **Voltage Measurement**, and also with the three-phase block **Three-Phase V-I Measurement**. Many blocks in the folders *Elements* and *Power Electronics* have the option *Measurements*. If this option is checked, a window appears in which the quantities of this block that are measured with the help of the block **Multimeter** have to be assigned. If to put **Multimeter** in the model, in which such blocks are present, and to open its window, it may be seen that this window consists of two parts. In the left part, a list appears containing all quantities that can be measured in this model. In order to select the quantities that have to be measured, they must be duplicated in the right part.

The folder *Control and Measurements* has a number of blocks, which are used for processing the main circuit quantities measured by the above-described devices. The outputs of these blocks are utilized for observation and control. These blocks are built by using Simulink blocks. The most important and often-utilized blocks are described briefly as follows.

The block **Fourier** performs Fourier analysis of the input signal over a running window of $T = 1/f$ width. The fundamental frequency f and the harmonic number, whose amplitude and phase are formed at the block outputs, are specified in the dialog box. The blocks **Mean** and **RMS** calculate the mean value or rms of the input signal over the same running window, respectively. For **RMS**, either the true value or the rms of the fundamental harmonic may be selected for computation.

The block **THD** computes the total value of distortion factor (THD) of the periodic signals as

$$THD = \frac{\sqrt{I_{rms}^2 - I_0^2 - I_1^2}}{I_1}, \qquad (1.39)$$

where I_{rms} is the rms value of the total input signal, I_1 is the rms of its fundamental harmonic, and I_0 is the direct component.

The block **Sequence Analyzer** calculates the positive, negative, and zero sequence components of the three-phase signal. The frequency, the harmonic number n ($n = 1$ for fundamental), and the type of the sequence—positive, negative, 0, or all three simultaneously—are specified in the dialog box. The vector of three-phase quantity enters the block. The block has two outputs: for the amplitude of the selected sequence and for its phase angle.

It is reasonable to use the block as a voltage feedback sensor for the voltage regulation tuned with $n = 1$ for positive sequence, and for the study of the method symmetric components, see example in Ref. [2].

A number of blocks are intended for power measurement. For instance, the block **Power** computes the instantaneous values of the active and reactive powers of a single-phase source at the fundamental frequency. The block **Power (3ph, Instantaneous)** uses the following relationships:

$$P = V_a I_a + V_b I_b + V_c I_c \qquad (1.40)$$

$$Q = \frac{1}{\sqrt{3}}(V_{bc} I_a + V_{ca} I_b + V_{ab} I_c) \qquad (1.41)$$

The block **Power (Positive-Sequence)** calculates the active and reactive powers of the positive-sequence fundamental harmonic in the unbalanced three-phase circuit, in which the voltages and the currents may have harmonics. The outputs are the fundamental harmonic amplitudes of the voltage and the current positive sequence and the quantities P and Q.

1.2 Advanced Devices and Units

In this section, more complicated models of devices and units are described, which are employed intensively in the modern RES. These units are developed partly in SimPowerSystems and partly by the author by using the SimPowerSystems blocks.

1.2.1 Batteries and Supercapacitors

Batteries and supercapacitors (SCs) store the electric energy when the power surplus is available in the system, to which they are connected, and give it back when the power shortage exists. In a certain sense, these devices are the complement of one another, because the battery has high energy density but low power density (under discharge, the battery voltage changes little, but the discharge current is rather little as well), and the SC, on the contrary, has high power density but low energy density (under discharge, the SC voltage decreases very quickly, but the discharge current is sufficiently large). Thus, the integration of SCs and batteries can form a rather perfect energy source. Of course, each device can operate independently.

The model of the **Battery** is included in SimPowerSystems that models in detail the charge–discharge processes in different types of batteries: lead acid, lithium ion, nickel–cadmium, and nickel metal hydride. The battery nominal voltage V_{nom}, its rated capacity Q_{rated} (Ah), the initial charge for simulation SOC (%), and the battery response time (30 s by default) are specified in the block dialog box. At this, the nominal voltage is the voltage value in the end of the discharge linear zone. For a lithium ion battery, consideration of the temperature effect can be selected.

A number of additional characteristics must be defined that can be done in two ways: if the option *Determined from the nominal parameters of the battery* is chosen, all the necessary parameters will be defined automatically; if this option is not checked, the necessary parameters have to be specified manually. These parameters are the maximum capacity Q_{max} (Ah) that is usually equal to 105% of the rated capacity; the fully charged voltage (V) V_{full}; the nominal discharge current, which the discharge characteristic is calculated for; the internal resistance that, for the model on default, is determined as the resistance, across which 1% of the power is dissipated: $R = 0.01V_{nom}/Q_{rated}$; the capacity under nominal voltage, which is the capacity that can be extracted from the battery until the voltage is kept not less than the nominal value; and exponential zone—the values V_{exp} and Q_{exp} correspond to the end of this zone. More detailed information about the described model is given in Ref. [1].

In RESs, the battery supplies to and is supplied from DC link of the system. At this, the battery voltage U_b and DC voltage U_{dc} are different; usually, $U_{dc} > U_b$. Therefore, the battery is connected to the DC link with the

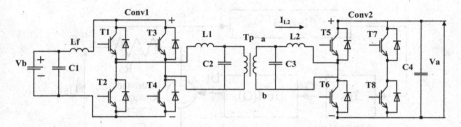

FIGURE 1.12
DC/DC converter with input–output electric separation.

boost/buck converter: when the power surplus exists, the battery is charged, till its $Q < Q_{max}$, the converter operates as the buck one; when the power shortage exists, the battery supplies DC link and discharges, the converter operates in the boost mode.

When the electric separation of the battery from DC link is not required, the boost/buck converter depicted in Figure 1.5 can be used. If such a separation is necessary, the diagram shown in Figure 1.12 can be utilized. Description of his circuits is given in Ref. [3].

SCs have a large capacitance—up to some thousand Farads in one unit that is obtained owing to a small spacing between the positive and negative charges. At this, however, a capacitor permissible voltage is low that necessitates their series connection. SC equivalent circuit is shown in Figure 1.13. The resistor R_p has effect on the slow discharge process that is not modeled further; therefore, the SC is modeled as a series connection R-C, where n_s and n_p are the numbers of series- and parallel-connected capacitor banks, respectively,

$$R = R_s \frac{n_s}{n_p}, \quad C = C_u \frac{n_p}{n_s}. \tag{1.42}$$

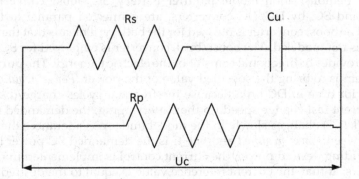

FIGURE 1.13
Supercapacitor simple equivalent circuit.

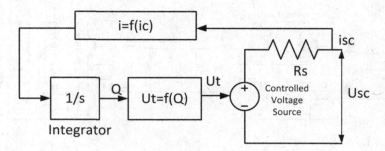

FIGURE 1.14
Block diagram of the supercapacitor model in SimPowerSystems.

Another model of a SC is included in SimPowerSystems. Its simplified structure is shown in Figure 1.14. The SC voltage

$$U_{sc} = U_t - R_s i_{sc}. \tag{1.43}$$

In turn, the internal voltage U_t is a function of the electric charge Q and a number of unit parameters, many of which have been determined from experimental tests, and they can be used as default values to represent a common SC. The expression for this function and parameter meanings are given in Ref. [1]. The electric charge is an integral of the internal current i that takes into account both the capacitor current i_{sc} and self-discharge phenomenon.

In the models of the simple systems, the model in Figure 1.13 operates faster. For example, if the process of capacitor charge from direct voltage through the resistor is modeled, one can see that such a process for the model in Figure 1.13 goes on twice faster than for the model in Figure 1.14. But with increasing model complexity, the difference decreases due to the effect of other units. For simulation, the SC model is chosen, depending on available data and model requirements.

In the demonstration model **parallel_battery_SC_boost_converter**, the battery and SC, by DC/DC converters, are connected parallel to the DC source. The boost converter is utilized for the battery; it means that the charge process is not modeled. The boost/buck converter is employed for SC; therefore, it provides bidirectional transfer of the electrical energy. The purpose of the system is to bring the specified value of the power (*Power_required*) that varies with time in DC link. Because the frequent cycles charge/discharge with current fast change speed up the battery aging, the demanded battery power (*Bat_C*) changes slowly in the model under consideration; the difference between *Power_required* and *Bat_C* is the demanded SC power (*SC_C*). For providing desired power, the current control is implemented in DC/DC converters; at that, the current reference value is equal to the desired power divided by the input converter voltage. The process is shown in Figure 1.15. The other examples of battery and SC utilization will be given further.

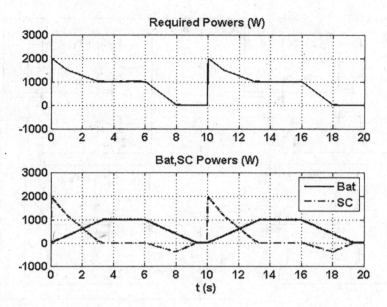

FIGURE 1.15
Processes in the demonstration model **parallel_battery_SC_boost_converter.**

1.2.2 Power Converters

The blocks of power electronics described in the previous section are commonly used in the modern RES. However, increasing a single-unit power promotes the utilization of more complicated converters, especially multilevel converters.

The cascaded H-Bridge multilevel converter consists of several single-phase bridges connected in series, as shown in Figure 1.16. Depending on the conducting semiconductors, each bridge can fabricate the voltage $\pm E$ or 0. Thus, the phase voltage can change from $-nE$ to nE, where n is the number of bridges in a phase; in this way, $2n + 1$ level, including 0, can be received. The rms value of the output phase-to-phase voltage is

$$U_l = 2kmnE, \tag{1.44}$$

where m is the modulation factor and $k = 0.612$. This scheme gives an opportunity to reach the high output voltage with the limited voltage of one bridge; the higher harmonic content in the output voltage decreases; the modular design makes production and maintenance lighter. The drawback is the necessity to have a lot of DC sources that are insulated from one another. A comparison of the reference harmonic signal having the demanded frequency with the triangle carrier waveform that has the frequency F_c can be used for the fabrication of the control pulses for the inverter. Both unipolar

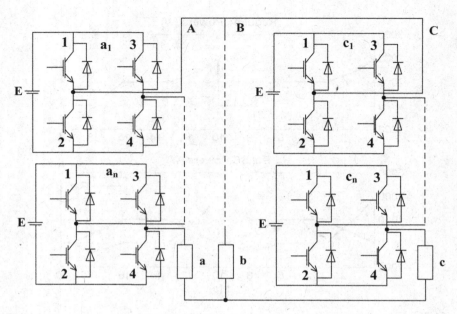

FIGURE 1.16
Cascaded H-Bridge multilevel converter.

and bipolar carrier waveforms can be utilized. At that, each bridge is controlled independently, and the carrier waveforms of the bridges are shifted one of the other by $2\pi/n$, in order to achieve the minimal THD value in the inverter phase-to-phase voltage. For unipolar modulation, carrier waveform for each IGBT changes in the interval (0 1) or (–1 0), and for bipolar modulation in the interval (–1 1). In the first case, the high harmonics are located around the frequency nF_c, and in the second case, around the frequency $2nF_c$ that makes filtering easier, but the average frequency of switching increases twofold.

The demonstration model of the cascaded inverter with five blocks in a phase is given in SimPowerSystems as **Five-Cell Multi-Level Converter**. Each single-phase bridge is supplied from the diode bridge, which, in turn, is powered from a separate transformer; the bipolar modulation is used. In RES, the use of the cascaded inverters is most reasonable in the photovoltaic systems, in which the parts of all the system can be used as the insulated DC sources, or in the STATCOM.

The models of the photovoltaic systems with the cascaded inverters are given further, models **Photo_Hinv1, 2a**.

As for STATCOM with the cascaded inverters, the examples of such devices are given in Ref. [3]. A supply source for each inverter bridge is a capacitor (the DC source is replaced with the capacitor in the diagram shown in Figure 1.16). Because STATCOM fabricates the reactive power, the average value of the voltage across the charged capacitor must not change in the ideal case. However, because of inevitable power losses, the control system has to have the circuits for capacitor voltage control. In the model **STATCOM_Hinv1**, Ref. [3], the cascaded multilevel inverter is used that was described earlier in this section.

STATCOM in **STATCOM_Hinv1** ensures the little content of the higher harmonic in the output voltage, but it is reached at the expense of the high frequency of switching-over inverter devices, which makes STATCOM employment difficult for the units of large power. The utilization of the selected harmonic elimination (SHE) method gives an opportunity to have the only switching over (switch on/switch off) of the each semiconductor for the period of the fundamental frequency. How it can be made, it is described in detail in Ref. [3] for 3 and 5 modules in a phase, the models **STATCOM_3_SHEM, STATCOM_3_SHEM**.

The special blocks for simulation of the cascaded inverters are introduced in the latest versions of SimPowerSystems. The block **Full-Bridge MMC** included in R 2016b version is intended for simulation of the STATCOM sets with such inverters. The converter consists of multiple series-connected power modules. Each power module consists of one H-Bridge and one capacitor on the DC side. The number of bridges, capacitance of the capacitors, and their initial voltage are specified in the block dialog box. So, the block models one inverter phase. The simulation modes are the same as for the blocks described earlier: *Switching devices, Switching function, and Average model.*

The block **PWM Generator (Multilevel)** for control of the described converter is developed that generates pulses for a PWM-controlled multilevel converter. The number N of the rectifier bridges controlled is given in the block dialog box. Four pulses for each bridge are fabricated; the bipolar modulation is used when the direct and inverse modulation signals are compared with the sawtooth waveform. The separate signals are produced for every bridge; they are shifted one about the other by $180/N$. The model **STATCOM (Detailed MMC Model with 22 Power Modules per Phase)** is included in SimPowerSystems, in which utilization of the above-mentioned blocks is demonstrated.

The model of the cascaded inverter **Full-Bridge MMC (External DC Links)** is developed in R 2017b version that is intended for simulation of the electrical drives, photovoltaic plants, and other sets with the separate supply of the different bridges. The number of feed points (+, –) is determined by the number of power modules pointed in the block dialog box. Of course, the use of these blocks simplifies the creation of complex models, but experience with their utilization shows that the simulation process slows down in this case. The above-mentioned models of the photovoltaic systems with the cascaded inverters **Photo_Hinv1, 2a** are made in several versions: for functioning in the old versions, for example, in R 2014a; for functioning in the last versions, for example, R 2019a, without use of the above-mentioned blocks phase, the letter N is added to the model name; with use of these blocks, NN is added.

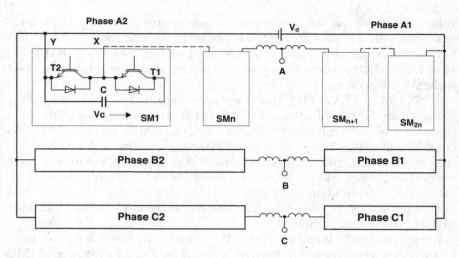

FIGURE 1.17
Modular multilevel converter.

At present, modular multilevel converters (MMC or M2C) are wide-spread in RES as inverters and rectifiers in the DC transmission lines and in STATCOM. The diagram of such three-phase converters is depicted in Figure 1.17. It consists of three same legs (phases) that are connected in parallel and are supplied from the DC source V_d. Each phase contains $2n$ same blocks (modules); the middle points of the legs are the converter outputs. Every module consists of two IGBTs with antiparallel diodes and the capacitor that is charged to the voltage V_c. The IGBTs are controlled in antiphase. When T_1 is turned on and T_2 is turned off, the voltage between the points X and Y is V_c; when T_2 is turned on and T_1 is turned off, this voltage is zero. Therefore, the voltage of a semi-phase (arm) V_{a1} or V_{a2} changes from 0 to nV_c. The value V_c is taken as V_d/n; the modules are controlled in such a way that at every moment, n modules are in the active state. Therefore, the arm voltage changes from 0 to V_d. If sinusoidal modulation is used, the reference voltages for the right and left arms of the phase A are

$$V_{ref\,1,2} = V_d/2 \pm mV_d/2 \sin \omega t, \qquad (1.45)$$

where $0 \le m \le 1$ is the modulation factor; for phases B and C, the reference voltages are shifted by 120° and 240°, respectively. The phase-to-phase voltage is

$$V_{ab} = \frac{V_d}{2} + m\frac{V_d}{2}\sin \omega t - \left[\frac{V_d}{2} + m\frac{V_d}{2}\sin\left(\omega t - \frac{2\pi}{3}\right)\right] = 0.866mV_d\sin\left(\omega t + \frac{\pi}{6}\right).$$

$$(1.46)$$

These converters consist of many same blocks that give an opportunity to build the high-voltage converter set, which often allows dispensing with the transformers for connection to the high-voltage grid. It is not difficult to set the additional blocks and, with this way, to increase the system reliability and survivability.

A converter control system has to provide the desired output voltage, the capacitor voltage balancing, and, as option, circulating current suppression (see further). One of the opportunities is to have PWM for every module, Ref. [6], but the large number of modules complicates the unit. Therefore, the modulation signal as in Equation (1.45) is usually used; after division by V_c, the result is rounded to the whole number, and in this way, the number of modules k is found, which must be in the active state. Such an operation repeats every $T_{me} = 1/F_c$ s.

Because of the most part of the modulation signal period $k < n$, the existence of the redundancy gives the possibility to use it for voltage balancing across the module capacitors. It is carried out in the following way.

It is supposed that first all the capacitors are charged to the voltage V_c. Afterward, if the phase current is positive, the capacitors of the modules that are in active state increase their voltage, and if the phase current is negative, they decrease their voltage. Therefore, if the current is positive, k modules are set in the active state that have the least voltage; if this current is negative, k modules are set in the active state that have the largest voltage.

With certain control system complication, the less THD value can be received even with a small module number. The integral part k = floor (V_{ref}/V_c) and the fractional part $k_{pwm} = rem(V_{ref}/V_c)$ are calculated; the value k determines how many modules must be activated in the next interval T_{me}. During interval $k_{pwm} T_{me}$, the module $k + 1$ is activated as well.

Figure 1.18 shows how this control system is implemented. Its kernel is the MATLAB function [**B**, **IX**] = sort (**V**$_c$), which is utilized to order the vector components. This function builds the vector **B** that consists of the vector **V**$_c$ components in an increasing order and the vector **IX** that contains the component indexes as they follow in the vector **B**. The switch **SWr** changes the index order for the reverse one; either **IX** or **IX**$_i$ are used depending on the current sign in the arm. The capacitor voltage vector **V**$_c$ and the value k are stored in the units for sampling and storage (**S/H**) during the time interval T_{me}.

Each block **Sort&Select** selects, from the input vector components, one that corresponds to the whole number that is at the input *CI*. When *CI* = 1, the first component is chosen, when *CI* = 2, the second one, and so on. Let, for instance, $k = 4$ and the least capacitor voltages are in the modules 5, 3, 9, and 7. In block 1, *CI* = 4, so that the output *Select* is 7; in block 2, *CI* = 3, so that the output *Select* is 9; in block 3, *CI* = 2, and the output *Select* is 3, the output *Select* of block 4 is 5. Because the outputs *Select* for blocks 5–10 are less than 1, the adder Sum2 is set in order to prevent the failure: this adder sends, in this case, to the comparison elements some number that is ineffective for comparison, for instance, 100.

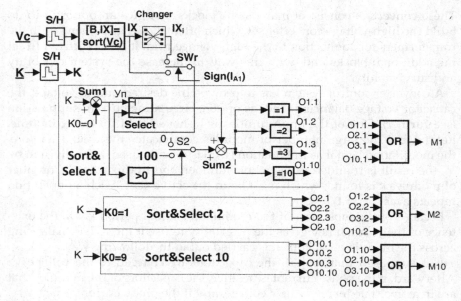

FIGURE 1.18
Block diagram of the modular multilevel converter control system.

Therefore, the logic 1 will be at the outputs O1.7, O2.9, O3.3, and O4.5 of the blocks 1, 2, 3, and 4, respectively, and the logic 0 at the outputs of the other blocks. With the help of the elements **OR**, the resulting signals are fabricated that control the modules M1–M10. It follows from what was said earlier that the gate pulses will be sent to the modules 3, 5, 7, and 9 having the least capacitor voltages.

Figure 1.19 shows how the module $k + 1$ is activated. On the fabrication of the next pulse T_{me}, the integrator is reset and afterward begins to integrate the signal that is equal to F_c. Until the integrator output $U_i < k_{pwm}$, the output

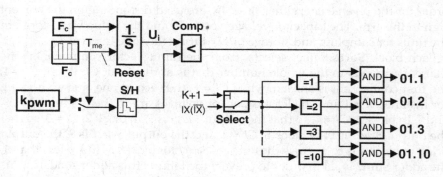

FIGURE 1.19
PWM for $k + 1$ module.

of the comparison block **Comp** = 1, and the module $k+1$ is in the active state. When $U_i > k_{pwm}$, **Comp** = 0, the module is deactivated. For the arm $A2$, the value k is a complement with respect to nine of the value k in the arm $A1$, and the module $k+1$ is activated after activation of the module $k+1$ in the arm $A1$ is over; in this way, the condition is kept that the number of active modules in a phase must be n always.

This inverter is used as the onshore inverter in the model **Wind_ PMSG_9M**, see the following Chapter 2. The special block for simulation of these converters is included in SimPowerSystems, **Half-Bridge MMC**, that models one arm of the MMC. The number of power modules in the arm, as well as the capacitance of the module capacitor and its initial voltage, has to be assigned in the block dialog box. The above-mentioned model is given in two versions: **Wind_PMSG_9M** with separate power blocks and **Wind_ PMSG_9Ma** with the block **Half-Bridge MMC**.

If MMC with PWM is modeled, the above-mentioned block **PWM Generator (Multilevel)** can be used in the mode *Half-bridge* converter. In this case, the number of pulses is $2N$ at the block output, where the number of blocks N in the arm is indicated in the block dialog box.

Let's go on with consideration of MMC. It may be written for the arm currents

$$I_+ = \frac{I_{dc}}{3} + i_a + i_{cir}, \qquad I_- = \frac{I_{dc}}{3} - i_a + i_{cir}, \qquad (1.47)$$

where I_{dc} is the DC link current, i_a is the load current (the output converter current), and i_{cir} is the circulating current in the converter leg, owing to capacitor voltage pulsations; this current does not manifest itself in the output converter current. Using Equation (1.46) and an equality of DC and AC powers, relation (1.48) may be obtained between DC current I_{dc} and the maximum I_{am} of the current i_a (supposing the reactive component is zero)

$$V_d I_{dc} = 1.5 I_{am} \times 0.5 V_d\, m, \qquad (1.48)$$

or

$$I_{dc} = 0.75 m I_{am}. \qquad (1.49)$$

From Equation (1.47)

$$i_{cir} = \frac{I_+ + I_-}{2} - \frac{1}{3} I_{dc}. \qquad (1.50)$$

The current i_{cir} contains a number of even harmonics, the second one is the largest; this current burdens the unit elements; therefore, it is reasonable to take steps for its decreasing, Refs. [7, 8].

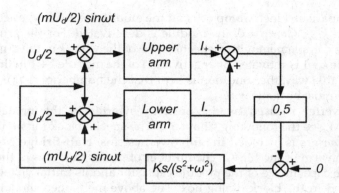

FIGURE 1.20
Controller structure for circulating current decrease.

The resonant controller is utilized for this purpose in Ref. [8], whose a transfer function is

$$G(s) = \frac{Ks}{s^2 + \omega_0^2},$$ (1.51)

where ω_0 is controller resonant frequency, taking as $4\pi \times 50$ to cancel the second harmonic. Such a controller achieves infinite gain at the AC frequency of ω_0 to force the steady-state error to zero.

It is accepted in the model **M2C_CC**, the module voltage is 1 kV, and $V_d = 10$ kV thereafter. Converters with these parameters were utilized in the models in Ref. [3]. The difference is that the subsystem **ResContr** is added that contains three controllers with the transfer function, Equation (1.51). The signals $\frac{I_+ + I_-}{2}$ of three phases come to the controller inputs; their outputs are added to the modulation signals with the proper sign. The controller structure for one phase is depicted in Figure 1.20.

It is supposed in simulation that the controllers are activated at $t = 1.5$ s. The process is displayed in Figure 1.21. One can see that the circulating current decreases essentially, its rms drops from 135 A to 42 A. The capacitors voltage pulsations decrease also a little.

1.2.3 Multiphase Electric Generators

1.2.3.1 Six-Phase IG

In Section 1.1.5, the standard models of the electric generators were considered, which are the three-phase ones. Most of the WG units of the small or medium power are equipped with just such generators. But at present, the

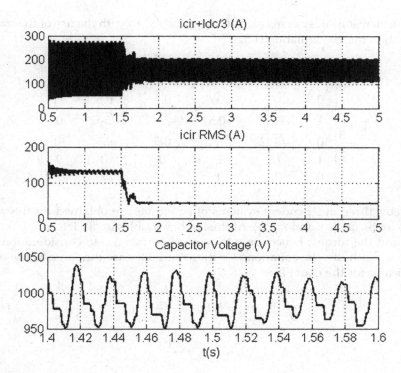

FIGURE 1.21
Processes in the model **M2C_CC**.

trend occurs to increase a WG single-unit power. The up-to-date WGs are provided with semiconductor devices, in order to increase their efficiency. At that, the power of the individual semiconductor devices is not sufficient, and they must be connected in series or (and) in parallel, which leads to some problems under the practical realization of such systems. More preferable is the converter parallel connection that gives an opportunity, with proper control, to decrease the content of the higher harmonics caused by the set. However, the stray currents appear in this case, which circulate between converters; these currents are caused by the difference of voltage instantaneous values and by the parameter spread. The utilization of the multiphase, for example, a six-phase generator, helps to exclude this problem.

At first, six-phase IG is considered. It has two systems of the three-phase stator windings designated as *abc* and *xyz*, having the mutual phase displacement of 30° (electrical); each three-phase system has its isolated neutral. These windings are placed in the same slots; therefore, a mutual inductance exists between three-phase windings. As it is proved in Refs. [9, 10], a six-dimension system of equations (for phases *A, B, C, X, Y, Z*) that describes the considered IG in the stationary reference frame can be replaced with three two-dimension subsystems (α, β), (μ_1, μ_2), and (z_1, z_2) with the use of the transformation matrix, Equation (1.52).

$$
T = \frac{1}{3}
\begin{vmatrix}
1 & -0.5 & -0.5 & \sqrt{3}/2 & -\sqrt{3}/2 & 0 \\
0 & \sqrt{3}/2 & -\sqrt{3}/2 & 0.5 & 0.5 & -1 \\
1 & -0.5 & -0.5 & -\sqrt{3}/2 & \sqrt{3}/2 & 0 \\
0 & -\sqrt{3}/2 & \sqrt{3}/2 & 0.5 & 0.5 & -1 \\
1 & 1 & 1 & 0 & 0 & 0 \\
0 & 0 & 0 & 1 & 1 & 1
\end{vmatrix}
\tag{1.52}
$$

Thereby, three independent systems of equations are obtained for the variables (α, β), (μ_1, μ_2), and (z_1, z_2). At this, only variables (α, β) determine the IG flux and the torque; hence, only they must be taken into consideration for system synthesis; the equations for these variables are the same as the ones applicable for the usual IG:

$$
v_{s\alpha} = R_s i_{s\alpha} + \frac{d\Psi_{s\alpha}}{dt}
\tag{1.53}
$$

$$
v_{s\beta} = R_s i_{s\beta} + \frac{d\Psi_{s\beta}}{dt}
\tag{1.54}
$$

$$
0 = R_r i_{r\alpha} + \frac{d\Psi_{r\alpha}}{dt} + \omega_r \Psi_{r\beta}
\tag{1.55}
$$

$$
0 = R_r i_{r\beta} + \frac{d\Psi_{r\beta}}{dt} - \omega_r \Psi_{r\alpha}
\tag{1.56}
$$

where ω_r is the rotor rotational speed.

The equations for the flux linkage are

$$
\begin{aligned}
\Psi_{s\alpha} &= (M + L_{ls}) i_{s\alpha} + M i_{r\alpha} \\
\Psi_{s\beta} &= (M + L_{ls}) i_{s\beta} + M i_{r\beta}
\end{aligned}
\tag{1.57}
$$

$$
\begin{aligned}
\Psi_{r\alpha} &= (M + L_{lr}) i_{r\alpha} + M i_{s\alpha} \\
\Psi_{r\beta} &= (M + L_{lr}) i_{r\beta} + M i_{s\beta},
\end{aligned}
\tag{1.58}
$$

where $M = 3L_{ms}$ are the resultant magnetizing inductance (L_{ms} is the stator magnetizing inductance).

The IG torque is

$$
T_e = 3Z_p \left(\Psi_{s\alpha} i_{s\beta} - \Psi_{s\beta} i_{s\alpha} \right).
\tag{1.59}
$$

Such a model was developed by the author in Ref. [3]. In this book, this model was updated; in particular, the model received a dialog box that would give an opportunity to dispense with a special program that had to be carried out before the model starts in the previous version. Remind that this model is based on the relationships given in Ref. [12]: the main flux linkages $\Psi_{m\alpha}$ and $\Psi_{m\beta}$ are introduced in Equations (1.25) and (1.26), which, in the above-taken notation, can be written as

$$\Psi_{m\alpha} = L_a \left(\frac{\Psi_{s\alpha}}{L_{ls}} + \frac{\Psi_{r\alpha}}{L_{lr}} \right) \tag{1.60}$$

$$\Psi_{m\beta} = L_a \left(\frac{\Psi_{s\beta}}{L_{ls}} + \frac{\Psi_{r\beta}}{L_{lr}} \right). \tag{1.61}$$

That gives the possibility to compute the generator currents as

$$i_{s\alpha} = \frac{\Psi_{s\alpha} - \Psi_{m\alpha}}{L_{ls}} \tag{1.62}$$

$$i_{s\beta} = \frac{\Psi_{s\beta} - \Psi_{m\beta}}{L_{ls}} \tag{1.63}$$

$$i_{r\alpha} = \frac{\Psi_{r\alpha} - \Psi_{m\alpha}}{L_{lr}} \tag{1.64}$$

$$i_{r\beta} = \frac{\Psi_{r\beta} - \Psi_{m\beta}}{L_{lr}} \tag{1.65}$$

and to use them in the voltage Equations (1.53)–(1.56).

This model takes into account saturation according to Ref. [12], for more details, see Ref. [3].

The currents $i_{\mu 1}$ and $i_{\mu 2}$ are found from equations

$$\begin{aligned} v_{\mu 1} &= (R_s + sL_{ls})i_{\mu 1} \\ v_{\mu 2} &= (R_s + sL_{ls})i_{\mu 2} \end{aligned} \tag{1.66}$$

where s is the Laplace operator symbol. The currents i_{z1} and i_{z2} are equal to zero for three-phase windings with insulated neutrals. Using these currents, with the help of the inverse transformation, the IG currents are calculated that IG sends to the external circuits or consumes from them. Five groups of the signals are collected at the output $m1$: three phase currents A, B, C; three phase currents X, Y, Z; two (in the axes α and β) stator flux linkages; rotor flux linkages; and rotor currents. The IG speed, its torque, and the rotor angle are collected at the output $m2$. All the quantities, except the latter, are given in pu. The base values of the equivalent three-phase IG of the same power are selected as the base ones.

In order to create the conditions for self-excitation, the initial values of the stator and rotor fluxes as the residual fluxes are assigned.

The model **IG_6PhN** demonstrates an example of employment of the developed six-phase IG model. On the whole, it is the model **IG_6Ph** from Ref. [3], in which the IG model is replaced with the new one. The IG is driven with the diesel. This set is intended for DC voltage fabrication. In principle, such a system unlikely has the practical sense and is given here in order to check the operation of the developed IG model.

The process of self-excitation is displayed in Figure 1.22, when the residual flux is about 2%. At the measured amplitude of the phase-to-phase IG voltage of 485 V, the rectified voltage must be 485 × 0.95 = 460 V; after simulation, we have the same value. Because the rectifiers have the individual loads of 40 Ω and the common load of 4 Ω, the load power is $P = 2 \times 460^2/40 + 920^2/4 = 214$ kW = 0.54 pu. On the second axis of the **Scope**, the

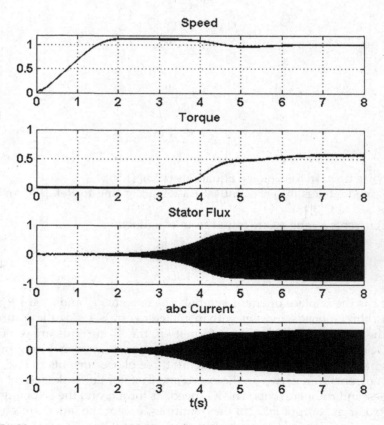

FIGURE 1.22
Process of IG self-excitation.

torque is 0.56. Since the rectified power of one winding is 107 kW, its input current is

$$I_r = \frac{1.07 \times 10^5 \times \sqrt{2}}{\sqrt{3} \times 485} = 180\,\text{A},$$

the capacitor current is $I_c = 485 \times 314 \times 2 \times 10^{-3}/\sqrt{2} = 215\,\text{A}$, therefore, the corresponding line current is $I_{cl} = 1.732 \times 215 = 372\,\text{A}$, so the phase IG current is $I_a = \sqrt{180^2 + 372^2} = 413\,\text{A}$. Because the rms value of the base current is $I_{brms} = 4 \times 10^5/1.732/400 = 577\,\text{A}$, I_a in pu is 0.72. On the first and second axes of the **Scope1**, this value is 0.73. One can see by **Scope1** that the mutual shift of the stator currents is 30°. This verifies the validity of the developed model and suitability of its utilization for RES simulation. In Chapter 2, this model is used for simulation of the WG, model **Wind_IG_6N**.

1.2.3.2 Six-Phase SG

The model of the six-phase SG is considered in this section. In this case, our results will be based on the results of Ref. [11]. Note that the demonstration model **power_6phsyncmachine** in Ref. [1] contains a model of such a generator, but this model does not take into account the leakage fluxes that encompass both winding systems, which can lead to erroneous results, especially when the generator is loaded by the electronic circuits with the forced-commutated semiconductors.

The model of the six-phase SG that considers the above-mentioned electromagnetic coupling of both windings was developed in Refs. [2, 3]; for this book, this model was updated: a saturation is taken into account and the dialog box is provided that makes it possible to dispense with the MATLAB program that had to be carried out before the model starts in the previous version of this model.

The winding magnetic coupling is determined by the parameters

$$L_{sm} = L_{ax} \cos g + L_{ay} \cos\left(g + 2\pi/3\right) + L_{az} \cos\left(g - 2\pi/3\right)$$

$$L_{sdq} = L_{ax} \sin g + L_{ay} \sin\left(g + 2\pi/3\right) + L_{az} \sin\left(g - 2\pi/3\right), g = \pi/6, \tag{1.67}$$

where $L_{ax}, L_{ay},$ and L_{az} are the mutual leakage inductances of the indicated phases of both windings. If to accept that the mutual inductance of the three-phase symmetrical windings is proportional to the cosine of angle between their axes, it follows from Equation (1.67) $L_{sm} \gg L_{sdq}$.

The flux linkage equations can be written in the reference frame that rotates with the rotor speed ω_r as

$$\frac{d\Psi_{qs1}}{dt} = U_{qs1} - \omega_r \Psi_{ds1} - R_s I_{q1} \tag{1.68}$$

$$\frac{d\Psi_{ds1}}{dt} = U_{ds1} + \omega_r \Psi_{qs1} - R_s I_{d1} \tag{1.69}$$

$$\frac{d\Psi_{qs2}}{dt} = U_{qs2} - \omega_r \Psi_{ds2} - R_s I_{q2} \tag{1.70}$$

$$\frac{d\Psi_{ds2}}{dt} = U_{ds2} + \omega_r \Psi_{qs2} - R_s I_{d2}. \tag{1.71}$$

$$\frac{d\Psi_{kq}}{dt} = -R_{kq} I_{kq} \tag{1.72}$$

$$\frac{d\Psi_{kd}}{dt} = -R_{kd} I_{kd} \tag{1.73}$$

$$\frac{d\Psi_f}{dt} = \frac{R_f}{L_{md}} V_f - R_f I_{fd}. \tag{1.74}$$

The index 1 is related to the *abc* windings; the index 2 is related to *xyz*. Equation (1.74) is written for reduced values, Ref. [12].

It is reasonable to introduce the flux linkages Ψ_{mq} and Ψ_{md} as

$$\Psi_{mq} = L_{mq}\left(I_{q1} + I_{q2} + I_{kq}\right) \tag{1.75}$$

$$\Psi_{md} = L_{md}\left(I_{d1} + I_{d2} + I_{kd} + I_{fd}\right), \tag{1.76}$$

where L_{mq} and L_{md} are the magnetizing inductances. It is reasonable also to use the parameters $k = L_{sm}/L_{sl}$, $b = L_{sdq}/L_{sl}$, and $d = 1 - k^2 - b^2$, where L_{sl} is the leakage inductance of the single winding. Then the following relationships for the stator currents can be obtained from the equations given in Ref. [11]:

$$I_{q1} = \frac{\Psi_{q1} - k\Psi_{q2} - (1-k)\Psi_{mq} + b(\Psi_{d2} - \Psi_{md})}{L_{sl}d} \tag{1.77}$$

$$I_{q2} = \frac{\Psi_{q2} - k\Psi_{q1} - (1-k)\Psi_{mq} - b(\Psi_{d1} - \Psi_{md})}{L_{sl}d} \tag{1.78}$$

$$I_{d1} = \frac{\Psi_{d1} - k\Psi_{d2} - (1-k)\Psi_{md} - b(\Psi_{q2} - \Psi_{mq})}{L_{sl}d} \tag{1.79}$$

$$I_{d2} = \frac{\Psi_{d2} - k\Psi_{d1} - (1-k)\Psi_{md} + b(\Psi_{q1} - \Psi_{mq})}{L_{sl}d} \tag{1.80}$$

$$I_{kq} = \frac{\Psi_{kq} - \Psi_{mq}}{L_{lkq}} \tag{1.81}$$

$$I_{kd} = \frac{\Psi_{kd} - \dot{\Psi}_{md}}{L_{lkd}} \quad (1.82)$$

$$I_{fd} = \frac{\Psi_f - \Psi_{md}}{L_{lf}}, \quad (1.83)$$

where L_{lkq}, L_{lkd}, and L_{lf} are the leakage inductances of the rotor appropriate windings.

After substitution of these formulas for currents in Equations (1.75) and (1.76), the relationships for Ψ_{mq} and Ψ_{md} are written as

$$\Psi_{mq} = L_{aq}\left[\frac{(\Psi_{q1} + \Psi_{q2})(1-k) + b(\Psi_{d2} - \Psi_{d1})}{dL_{sl}} + \frac{\Psi_{kq}}{L_{lkq}}\right] \quad (1.84)$$

$$\Psi_{md} = L_{ad}\left[\frac{(\Psi_{d1} + \Psi_{d2})(1-k) + b(\Psi_{q2} - \Psi_{q1})}{dL_{sl}} + \frac{\Psi_{kd}}{L_{lkd}} + \frac{\Psi_f}{L_{lf}}\right] \quad (1.85)$$

$$L_{aq} = \left[\frac{1}{L_{mq}} + \frac{1}{L_{lkq}} + \frac{2(1-k)}{dL_{sl}}\right]^{-1} \quad (1.86)$$

$$L_{ad} = \left[\frac{1}{L_{md}} + \frac{1}{L_{lkd}} + \frac{1}{L_{lf}} + \frac{2(1-k)}{dL_{sl}}\right]^{-1}. \quad (1.87)$$

The SG torque is

$$T_e = (I_{q1} + I_{q2})\Psi_{md} - (I_{d1} + I_{d2})\Psi_{mq}. \quad (1.88)$$

The electrical part of the six-phase SG model consists of seven subsystems. The subsystem $\mathbf{V_{abc}}$–$\mathbf{V_{dq}}$ calculates the voltages U_{ds1}, U_{ds2}, U_{qs1}, U_{qs2} with the use of the measured linear voltages of the stator windings at the known position of the rotor. The subsystem T_e calculates the torque by Equation (1.88). The subsystems $\mathbf{Iqd\_abc}$ and $\mathbf{Iqd\_xyz}$, with the known position of the rotor, calculate the currents of the stator windings, both in pu and in the absolute values, for connection with the external circuits. The active and reactive powers of the generator are calculated in the subsystem $\mathbf{PQ}$, and the rotor torque angle δ in the subsystem $\mathbf{Sub\_delta}$.

The main computations are carried out in the subsystem $\mathbf{Stator\_Rotor}$. It consists of seven subsystems, in which Equations (1.77)–(1.83) are solved, and includes the subsystem $\mathbf{SM\_Flux}$ that consists of two subsystems $\mathbf{SM\_Flux\_q}$ and $\mathbf{SM\_Flux\_d}$, having a similar structure, in which both the currents and the flux linkages Ψ_{md} and Ψ_{mq} are calculated. The saturation computations are fulfilled in this subsystem as well.

As in the library model of a synchronous machine, saturation computations are performed separately for the d and q axes; for the latter, this is done only for the round rotor machine, that is, for $L_{md} \approx L_{mq}$. Like for the IG, the saturation curve is defined by the $2 \times n$ matrix where the first row defines the stator currents $i(k)$, and the second one defines the stator voltages $v(k)$ under these currents in the no-load condition in pu; n is the number of points of the magnetization curve. The first point $[i(k), v(k)]$ is the one where the effect of saturation begins to manifest itself. It is assumed that the values of L_{md} and L_{mq} indicated in the dialog box correspond to the initial, unsaturated part of the curve, and that the relative values of v and Ψ_{md} coincide. Then for $\Psi_{md} < v(1)$ $L_{mdsat} = L_{md}$, and for $\Psi_{md} > v(1)$ $L_{mdsat} = L_{md} \dfrac{i(1)}{v(1)} \dfrac{v}{i}$. This value is used to calculate L_{ad} and Ψ_{md}. The calculation of L_{mqsat} under $L_{md} \approx L_{mq}$ is made similarly. It is possible to exclude calculations related to saturation.

The model **SG_6PhN** is the model **IG_6PhN** considered earlier, in which six-phase IG is replaced with the six-phase SG. Smoothing reactors are installed in the rectifier supply circuits, parallel capacitors are excluded. The main parameters of the generator are retained: 400 kW and 400 V. The remaining parameters can be seen in the dialog box of the generator. To excite the generator, the model of the ST2A exciter is used, described, for example, in Ref. [3]. The acceleration process is simulated from zero speed to nominal with zero initial conditions. The resulting process is shown in Figure 1.23. The active power dissipated by the load resistances at $U_{dc} = 920$ V and by the ballast resistances is 240 kW or 0.6 pu, which is agree with the data in Figure 1.23. In Chapter 2, this SG is used in the model **Wind_SG_2aN**.

1.2.3.3 Nine-Phase SG

Owing to an increase in WG power, especially for the offshore sets, an increase in the number of SG phases becomes an actual task. Nine-phase SG has three three-phase stator windings, the axes of which are shifted relative to each other by 20° (electrical). We denote these windings as *abc*, *xyz*, *uvw* (Figure 1.24). The rotor windings are usual. Deriving the equations of such a SG, the same technique is used as for the equations of a six-phase SG, Ref. [11].

Let $\mathbf{I}_a, \mathbf{I}_x, \mathbf{I}_u, \mathbf{I}_r$ are the three-dimensional current vectors where $\mathbf{I}_a = [i_a, i_b, i_c]^T$ and the same for $\mathbf{I}_x, \mathbf{I}_u, \mathbf{I}_r = [i_{kd}, i_{kq}, i_f]^T$; $\mathbf{I} = [\mathbf{I}_a, \mathbf{I}_x, \mathbf{I}_u, \mathbf{I}_r]^T$ is 12-dimensional vector, and the vectors of the voltages $\mathbf{U}$ and the flux linkages $\mathbf{\Psi}$ are defined in the same way. The inductance matrix $\mathbf{L}$ that links the flux linkages with the currents, as in Equation (1.23), can be represented as the sum of two matrices, one of which takes into account leakage fluxes, and the second is related to the main generator flux. The former consists of two submatrices, which take into account the leakage fluxes of the stator and rotor windings, respectively. The latter is the same as in the previous SG models.

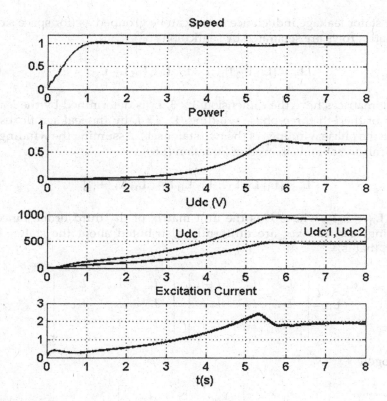

FIGURE 1.23
Acceleration process of six-phase SG with diesel generator.

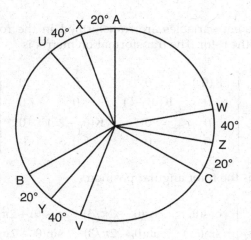

FIGURE 1.24
Arrangement of the winding in nine-phase SG.

The stator leakage inductance terms can be grouped as (for space saving, the matrix rows are separated by semicolon):

$$\mathbf{L_s} = [\mathbf{L}_{11}\ \mathbf{L}_{12}\ \mathbf{L}_{13}; \mathbf{L}_{21}\ \mathbf{L}_{22}\ \mathbf{L}_{23}; \mathbf{L}_{31}\ \mathbf{L}_{32}\ \mathbf{L}_{33}]\qquad(1.89)$$

All submatrices have the dimension 3×3, $\mathbf{L}_{ii}$ is determined by the leakage fluxes of the i-th three-phase winding, $\mathbf{L}_{ij}$, $i \neq j$, by the leakage fluxes that couple the phase windings of the two stator sets. Assuming the windings are identical, one obtains instead of Equation (1.89)

$$\mathbf{L_s} = [\mathbf{L}_{11}\ \mathbf{L}_{12}\ \mathbf{L}_{13}; \mathbf{L}_{12}^T\ \mathbf{L}_{11}\ \mathbf{L}_{12}; \mathbf{L}_{13}^T\ \mathbf{L}_{12}^T\ \mathbf{L}_{11}].\qquad(1.90)$$

Here $\mathbf{L}_{11} = L_{sl}\mathbf{1}_3$, where $\mathbf{1}_3$ is the unit matrix of the third order. Since the windings *abc* and *xyz* are uniformly distributed about the stator, $\mathbf{L}_{12}$ is cyclic, that is,

$$\mathbf{L}_{12} = \begin{bmatrix} L_{ax} & L_{ay} & L_{az} \\ L_{bx} & L_{by} & L_{bz} \\ L_{cx} & L_{cy} & L_{cz} \end{bmatrix} = \begin{bmatrix} L_{ax} & L_{ay} & L_{az} \\ L_{az} & L_{ax} & L_{ay} \\ L_{ay} & L_{az} & L_{ax} \end{bmatrix}.\qquad(1.91)$$

Analogous

$$\mathbf{L}_{13} = \begin{bmatrix} L_{au} & L_{av} & L_{aw} \\ L_{aw} & L_{au} & L_{av} \\ L_{av} & L_{aw} & L_{au} \end{bmatrix}.\qquad(1.92)$$

As usual, the system variables are transformed in the rotating reference frame relating to the rotor. The transformation matrix is

$$\mathbf{T} = \begin{bmatrix} \mathbf{K}(\theta_r) & 0 & 0 & 0 \\ 0 & \mathbf{K}(\theta_r - \zeta) & 0 & 0 \\ 0 & 0 & \mathbf{K}(\theta_r - 2\zeta) & 0 \\ 0 & 0 & 0 & \mathbf{1}_3 \end{bmatrix},\qquad(1.93)$$

where $\zeta = \pi/9$, θ_r is the rotor angular position,

$$\mathbf{K}(\theta_r) = \frac{2}{3} \begin{bmatrix} \cos(\theta_r) & \cos(\theta_r - 2\pi/3) & \cos(\theta_r + 2\pi/3) \\ \sin(\theta_r) & \sin(\theta_r - 2\pi/3) & \sin(\theta_r + 2\pi/3) \\ 1/\sqrt{2} & 1/\sqrt{2} & 1/\sqrt{2} \end{bmatrix}.\qquad(1.94)$$

Transformation from the stationary reference frame *abc* to the rotating one *dq* is carried out by relationships $\Psi_{dq} = \mathbf{T}\Psi_{abc} = \mathbf{T}\mathbf{L}\mathbf{I}_{abc} = \mathbf{T}\mathbf{L}\mathbf{T}^{-1}\mathbf{T}\mathbf{I}_{abc} = \mathbf{L}_{dq}\mathbf{I}_{dq}$, $\mathbf{L}_{dq} = \mathbf{T}\mathbf{L}\mathbf{T}^{-1}$. Applying this transformation to the matrix $\mathbf{L}_s$ [using the top 3×3 block from Equation (1.93)], we obtain

$$
L_{sdq} = \begin{bmatrix}
L_{sl} & 0 & L_{m1} & -L_{dq1} & L_{m2} & -L_{dq2} \\
0 & L_{sl} & L_{dq1} & L_{m1} & L_{dq2} & L_{m2} \\
L_{m1} & L_{dq1} & L_{sl} & 0 & L_{m1} & -L_{dq1} \\
-L_{dq1} & L_{m1} & 0 & L_{sl} & L_{dq1} & L_{m1} \\
L_{m2} & L_{dq2} & L_{m1} & L_{dq1} & L_{sl} & 0 \\
-L_{dq2} & L_{m2} & -L_{dq1} & L_{m1} & 0 & L_{sl}
\end{bmatrix}, \tag{1.95}
$$

where $L_{m1} = L_{ax} \times \cos \zeta + L_{ay} \times \cos (\zeta + 2pi/3) + L_{az} \times \cos (\zeta - 2pi/3)$; $L_{dq1} = L_{ax} \times \sin \zeta + L_{ay} \times \sin (\zeta + 2pi/3) + L_{az} \times \sin (\zeta - 2pi/3)$; $L_{m2} = L_{m1}$ and $L_{dq2} = L_{dq1}$ with replacing ζ for 2ζ and L_{ax}, L_{ay}, L_{az} for L_{au}, L_{av}, L_{aw}, respectively. The rows corresponding to i_0 are omitted in the matrix Equation (1.95), since it is assumed that all three stator windings have their own insulated neutrals, so the sum of the phase currents is zero.

Thus, taking into account the terms associated with the main flux and the leakage fluxes of the rotor windings, as in Equation (1.30), one can write in the reference frame that rotates with the rotor speed (by assumption, $L_{mq} \approx L_{md}$):

$$\Psi_{qs1} = L_{sl}I_{q1} + L_{m1}I_{q2} - L_{dq1}I_{d2} + L_{m2}I_{q3} - L_{dq2}I_{d3} + L_{mq}\left(I_{q1} + I_{q2} + I_{q3} + I_{kq}\right)$$

$$\Psi_{ds1} = L_{sl}I_{d1} + L_{dq1}I_{q2} + L_{m1}I_{d2} + L_{dq2}I_{q3} + L_{m2}I_{d3} + L_{md}\left(I_{d1} + I_{d2} + I_{d3} + I_{kd} + I_f\right)$$

$$\Psi_{qs2} = L_{m1}I_{q1} + L_{dq1}I_{d1} + L_{sl}I_{q2} + L_{m1}I_{q3} - L_{dq1}I_{d3} + L_{mq}\left(I_{q1} + I_{q2} + I_{q3} + I_{kq}\right)$$

$$\Psi_{ds2} = -L_{dq1}I_{q1} + L_{m1}I_{d1} + L_{sl}I_{d2} + L_{dq1}I_{q3} + L_{m1}I_{d3} + L_{md}\left(I_{d1} + I_{d2} + I_{d3} + I_{kd} + I_f\right)$$

$$\Psi_{qs3} = L_{m2}I_{q1} + L_{dq2}I_{d1} + L_{m1}I_{q2} + L_{dq1}I_{d2} + L_{sl}I_{q3} + L_{mq}\left(I_{q1} + I_{q2} + I_{q3} + I_{kq}\right)$$

$$\Psi_{ds3} = -L_{dq2}I_{q1} + L_{m1}I_{d1} - L_{dq1}I_{q2} + L_{m1}I_{d2} + L_{sl}I_{d3} + L_{md}\left(I_{d1} + I_{d2} + I_{d3} + I_{kd} + I_f\right)$$

$$\Psi_{kq} = L_{lkq}I_{kq} + L_{mq}\left(I_{q1} + I_{q2} + I_{q3} + I_{kq}\right)$$

$$\Psi_f = L_{lf}I_f + L_{md}\left(I_{d1} + I_{d2} + I_{d3} + I_{kd} + I_f\right)$$

$$\Psi_{kd} = L_{lkd}I_{kd} + L_{md}\left(I_{d1} + I_{d2} + I_{d3} + I_{kd} + I_f\right).$$

$$\tag{1.96}$$

The voltage equations are Equations (1.68)–(1.74) with addition

$$\frac{d\Psi_{qs3}}{dt} = U_{qs3} - \omega_r \Psi_{ds3} - R_s I_{q3} \tag{1.97}$$

$$\frac{d\Psi_{ds3}}{dt} = U_{ds3} + \omega_r \Psi_{qs3} - R_s I_{d3}. \tag{1.98}$$

Using Equations (1.68)–(1.74) and Equations (1.96)–(1.98), the SG currents and flux linkages can be found and SG torque can be calculated as

$$T_e = 1.5 Z_p \left[\left(I_{q1} + I_{q2} + I_{q3} \right) \Psi_{md} - \left(I_{d1} + I_{d2} + I_{d3} \right) \Psi_{mq} \right] \tag{1.99}$$

$$\Psi_{mq} = L_{mq} \left(I_{q1} + I_{q2} + I_{q3} + I_{kq} \right) \tag{1.100}$$

$$\Psi_{md} = L_{md} \left(I_{d1} + I_{d2} + I_{d3} + I_{kd} + I_f \right) \tag{1.101}$$

Here the indices 1, 2, 3 indicate the windings *abc*, *xyz*, *uvw*, respectively.

As shown in Ref. [13], the mutual inductance of stator windings is the sum of two terms, one of which is determined by the number of sharing stator slots of the different windings, and the second is determined by the angular distance between the axes of the windings. The calculation or experimental determination of these terms is a rather complicated task; at the same time, the data given in the cited literature show that the first term is much less than the second. Therefore, we further assume that the mutual inductances of the windings are $2/3\, k\, L_{sl} \cos \gamma$, where k is the proportionality factor and γ is the angle between the winding axes. Then we have $L_{ax} = 2/3kL_{sl} \cos \zeta$, $L_{ay} = 2/3kL_{sl} \cos(\zeta + 2\pi/3)$, $L_{az} = 2/3kL_{sl} \cos(\zeta - 2\pi/3)$, so that $L_{m1} = 2/3kL_{sl} [\cos^2 \zeta + \cos^2(\zeta + 2\pi/3) + \cos^2(\zeta + 2\pi/3)] = kL_{sl}$, $L_{dq1} = 0.67kL_{sl}[\cos \zeta \times \sin \zeta + \cos(\zeta + 2\pi/3) \times \sin(\zeta + 2\pi/3) + \cos(\zeta - 2\pi/3) \times \sin(\zeta - 2\pi/3)] = 0$, $L_{au} = 2/3kL_{sl} \cos 2\zeta$, $L_{ay} = 2/3kL_{sl} \cos(2\zeta + 2\pi/3)$, $L_{az} = 2/3kL_{sl} \cos(2\zeta - 2\pi/3)$, therefore, $L_{m2} = k\, L_{sl}$, $L_{dq2} = 0$. Thus, the flux linkages along the d and q axes are independent and

$$\Psi_{qs1} = L_{sl} \left(I_{q1} + k I_{q2} + k I_{q3} \right) + \Psi_{mq}$$

$$\Psi_{ds1} = L_{sl} \left(I_{d1} + k I_{d2} + k I_{d3} \right) + \Psi_{md}$$

$$\Psi_{qs2} = L_{sl} \left(k I_{q1} + I_{q2} + k I_{q3} \right) + \Psi_{mq}$$

$$\Psi_{ds2} = L_{sl} \left(k I_{d1} + I_{d2} + k I_{d3} \right) + \Psi_{md} \tag{1.102}$$

$$\Psi_{qs3} = L_{sl} \left(k I_{q1} + k I_{q2} + I_{q3} \right) + \Psi_{mq}$$

$$\Psi_{ds3} = L_{sl} (k I_{d1} + k I_{d2} + I_{d3}) + \Psi_{md},$$

hence it follows

$$I_{q1} = \frac{F_1 + k(F_1 - F_2 - F_3)}{dL_{sl}} \qquad (1.103)$$

$$I_{q2} = \frac{F_2 - k(F_1 - F_2 + F_3)}{dL_{sl}} \qquad (1.104)$$

$$I_{q3} = \frac{F_3 - k(F_1 + F_2 - F_3)}{dL_{sl}} \qquad (1.105)$$

$$F_i = \Psi_{qsi} - \Psi_{mq}, d = (1-k)(1+2k),$$

and the same expressions for the currents I_{d1}, I_{d2}, and I_{d3}. After substitution of these expressions for currents in Equations (1.100) and (1.101), taking in mind Equations (1.96), it turns out

$$\Psi_{mq} = L_{aq} \left[\frac{\Psi_{kq}}{L_{lkq}} + \frac{(\Psi_{q1} + \Psi_{q2} + \Psi_{q3})}{(1+2k)L_{sl}} \right] \qquad (1.106)$$

$$L_{aq} = \left[\frac{1}{L_{mq}} + \frac{1}{L_{lkq}} + \frac{3}{(1+2k)L_{sl}} \right]^{-1}. \qquad (1.107)$$

Analogous for Ψ_{md}

$$\Psi_{md} = L_{ad} \left[\frac{\Psi_f}{L_{lf}} + \frac{\Psi_{kd}}{L_{lkd}} + \frac{(\Psi_{d1} + \Psi_{d2} + \Psi_{d3})}{(1+2k)L_{sl}} \right] \qquad (1.108)$$

$$L_{ad} = \left[\frac{1}{L_{md}} + \frac{1}{L_{lf}} + \frac{1}{L_{lkd}} + \frac{3}{(1+2k)L_{sl}} \right]^{-1}. \qquad (1.109)$$

The model **SM_9_PHASE** is developed with utilization of the above-given relationships. It is similar in structure to the six-phase SG model described above. The saturation of SG is taken into account in the same way as in the model of a six-phase generator. The SG parameters are specified in model dialog box. Unlike the six-phase SG, the values of resistances and inductances, as well as the moment of inertia, are given in the SI units; their conversion into pu is carried out according to the instructions in the *Initialization* pane of the model dialog box when the model starts. The initial values of velocity and flux linkages, as well as the saturation characteristic, are given in pu.

In the model **SG_9ph**, designed to test the model adequacy, the prime mover with its speed controller is simply modeled as a PI controller; the SG with the power of 8 MW and the voltage of each winding 2400 V is loaded with three three-phase resistors with the power of 2 MW each. Excitation is carried out from the exciter ST2A; the average value of the voltage of the three loads is regulated. If to run simulation, it can be verified that the amplitude of the load phase voltage is 1960 V, which corresponds to the rms line voltage 2400 V; the voltage waveforms of the individual loads are shifted relative to each other by 20°. The values of power and torque are 0.75 (3 × 2/8 MW). The amplitude of the phase current must be equal to $I_a = \sqrt{2} \times 2 \times 10^6 / \sqrt{3} \times 2400 = 680$ A, which, with the base current of $I_b = 2 \times 8 \times 10^6 / (3 \times 1960) = 2721$ A, gives 0.25 pu. The oscilloscope **Scope Currents** really shows this value of the phase current amplitude. The load angle δ at the nominal (base frequency) can be found from the ratio Equation (1.110), Ref. [12]:

$$T_e = \frac{I_f L_{md} V_s}{(L_{md} + L_{sle})} \sin \delta, \qquad (1.110)$$

all the values are in pu, V_s is the stator voltage. In the case under consideration, $V_s = 1$, $T_e = 0.75$, L_{sle} is the leakage equivalent inductance of three stator windings connected in parallel. Taking with some assumptions $L_{sle} = L_{sl}/3$ and observing by the **Scope 3** that the steady-state value $I_f = 1.22$, we find

$$\delta = \arcsin \left[\frac{0.75 \times (2.3 + 0.27/3)}{1.22 \times 2.3} \right] = 39.7°;$$

according to the display **Delta** δ = 40°. The aforesaid proves the adequacy of the developed model of the nine-phase SG. In Chapter 2, this model is applied in the model **Wind_SG_4**.

Along with the above model, a second version of the nine-phase SG model, **SM_9_PHASE_M**, was developed, which takes into account the possible asymmetry of the windings and their mutual influence along the d and q axes. This model is used in the model **SG_9phN** and functions in accordance with relations (1.22) and (1.23), where in this case **V**, **Ψ**, and **I** are nine-dimensional vectors, **L**, **R**, **W** are the square matrices 9 × 9, **V** = $[U_{qs1}\ U_{ds1}\ U_{qs2}\ U_{ds2}\ U_{qs3}\ U_{ds1}\ 0\ 0\ U_f]$, where the first six components are the stator voltage components, U_f is the excitation voltage, **Ψ** = $[\Psi_{qs1}\ \Psi_{ds1}\ \Psi_{qs2}\ \Psi_{ds2}\ \Psi_{qs3}\ \Psi_{ds3}\ \Psi_{kq}\ \Psi_{kd}\ \Psi_f]$, the vector **I** is ordered likewise, **R** is the diagonal matrix of the resistances, and **W** is the matrix that consists of zero elements except $w(1, 2) = w(3, 4) = w(5, 6) = \omega_r$, $w(2, 1) = w(4, 3) = w(6, 5) = -\omega_r$,

$$\mathbf{L} = \Big[\big(L_{sl} + L_{mq}\big), 0, \big(L_{m1} + L_{mq}\big), -L_{dq1}, \big(L_{m2} + L_{mq}\big), -L_{dq2}, L_{mq}, 0, 0;$$

$$0, \big(L_{sl} + L_{md}\big), L_{dq1}, \big(L_{m1} + L_{md}\big), L_{dq2}, \big(L_{m2} + L_{md}\big), 0, L_{md}, L_{md};$$

$$\big(L_{m1} + L_{mq}\big), L_{dq1}, \big(L_{sl} + L_{mq}\big), 0, \big(L_{m1} + L_{mq}\big), -L_{dq1}, L_{mq}, 0, 0;$$

$$-L_{dq1}, \big(L_{m1} + L_{md}\big), 0, \big(L_{sl} + L_{md}\big), L_{dq1}, \big(L_{m1} + L_{md}\big), 0, L_{md}, L_{md};$$

$$\big(L_{m2} + L_{mq}\big), L_{dq2}, \big(L_{m1} + L_{mq}\big), L_{dq1}, \big(L_{sl} + L_{mq}\big), 0, L_{mq}, 0, 0; \qquad (1.111)$$

$$-L_{dq2}, \big(L_{m1} + L_{md}\big), -L_{dq1}, \big(L_{m1} + L_{md}\big), 0, \big(L_{sl} + L_{md}\big), 0, L_{md}, L_{md};$$

$$L_{mq}, 0, L_{mq}, 0, L_{mq}, 0, \big(L_{mq} + L_{lkq}\big), 0, 0;$$

$$0, L_{md}, 0, L_{md}, 0, L_{md}, 0, \big(L_{md} + L_{lkd}\big), L_{md};$$

$$0, L_{md}, 0, L_{md}, 0, L_{md}, 0, L_{md}, \big(L_{lf} + L_{md}\big)\Big].$$

The electromagnetic torque is computed as

$$T_e = L_{md}\big(I_{qs1} + I_{qs2} + I_{qs3}\big)\big(I_{ds1} + I_{ds2} + I_{ds3} + I_{kd} + I_f\big)$$
$$- L_{mq}\big(I_{ds1} + I_{ds2} + I_{ds3}\big)\big(I_{qs1} + I_{qs2} + I_{qs3} + I_{kq}\big). \qquad (1.112)$$

In the developed model, the input and output blocks of the previous model are retained. Saturation is not taken into account, since, in this case, it would be necessary to invert the time-variable inductance matrix 9 × 9 at each calculation step that would significantly increase the simulation time. In the dialog box, the values of the coefficients are given that determine the values of inductances $L_{m1}, L_{m2}, L_{dq1}, L_{dq2}$: $L_{m1} = k_{m1}L_{sl}$ and so on.

It can be seen by running the model **SG_9phN** that the instability of the system occurs for some values of these coefficients. For example, when $k_{m1} = 0.66$, $k_{m2} = 0.6$, instability occurs when $k_{dq1} = k_{dq2} = 0.46$. Investigation of the eigenvalues of the matrix $\mathbf{W} + \mathbf{RL}^{-1}$ reveals that with $k_{dq1} = k_{dq2} = 0.5$, one of the eigenvalues changes its sign; the discrepancy between the two values of the coefficients is due to the fact that the eigenvalues of the matrix determine the stability boundary of the continuous system, while the system is modeled as discrete.

The influence of the magnetic coupling of the windings becomes more noticeable when they are loaded by the switched semiconductor devices.

1.2.3.4 Six-Phase PMSG

PMSG with two windings that are mutually shifted by 30° are employed in the same cases as the six-phase SG with wound rotor. The voltage equations

are the same as Equations (1.68) through (1.71); the equations that link flux
linkages and currents may be written as

$$\begin{aligned}
\Psi_{q1} &= L_q I_{q1} + L_{sm} I_{q2} - L_{sdq} I_{d2} \\
\Psi_{d1} &= L_d I_{d1} + L_{sdq} I_{q2} + L_{sm} I_{d2} + \Psi_r \\
\Psi_{q2} &= L_{sm} I_{q1} + L_{sdq} I_{d1} + L_q I_{q2} \\
\Psi_{d2} &= -L_{sdq} I_{q1} + L_{sm} I_{d1} + L_d I_{d2} + \Psi_r.
\end{aligned}$$

(1.113)

The inductances L_{sm} and L_{sdq} are determined by Equation (1.67). Let
$\Psi_r = |0 \quad \Psi_r \quad 0 \quad \Psi_r|^T$. Then the voltage equations can be written as

$$\frac{d\Psi}{dt} = \mathbf{V} - \mathbf{RI} - \mathbf{W}\Psi,$$

(1.114)

where, in the case, $\mathbf{V}$, Ψ, and $\mathbf{I}$ are four-dimensional vectors, $\mathbf{R}$ and $\mathbf{W}$ are the
square matrices 4×4, $\mathbf{V} = \begin{bmatrix} U_{qs1} & U_{ds1} & U_{qs2} & U_{ds2} \end{bmatrix}^T$, the vectors Ψ and $\mathbf{I}$ are
ordered alike. $\mathbf{R}$ is the diagonal matrix of the resistances, and $\mathbf{W}$ is the matrix
that consists of zero elements except $w(1, 2) = w(3, 4) = \omega_r$, $w(2, 1) = w(4, 3) = -\omega_r$. The relationships (1.113) is written as

$$\Psi - \Psi_r = \mathbf{LI},$$

(1.115)

where $\mathbf{L} = \begin{vmatrix} L_q & 0 & L_{sm} & -L_{sdq} \\ 0 & L_q & L_{sdq} & L_{sm} \\ L_{sm} & L_{sdq} & L_q & 0 \\ -L_{sdq} & L_{sm} & 0 & L_q \end{vmatrix}.$

(1.116)

The generator electromagnetic torque is

$$T_e = 1.5 Z_p \left(\Psi_{d1} I_{q1} - \Psi_{q1} I_{d1} + \Psi_{d2} I_{q2} - \Psi_{q2} I_{d2} \right).$$

(1.117)

It is possible to prove by using Equations (1.115) and (1.116) that in pu

$$T_e = \Psi_r \left(I_{q1} + I_{q2} \right) + \left(L_d - L_q \right) \left(I_{q1} I_{d1} + I_{d2} I_{q2} \right).$$

(1.118)

Equations (1.114) and (1.115) are solved together in the developed model.
Unlike the library model, the calculations are performed in pu, as in the
models of other generators. In the model dialog box, the generator parameters are entered, including its power that is not required for the library
model. The structure of the model is basically the same as the model of the
nine-phase SG (with a decrease in the number of inputs and outputs). The
difference is in the structure of the **Main Subsystem**. The output m gives

the opportunity to measure three phase currents I_{abc}, three phase currents I_{xzy}, four currents I_{qd}, active and reactive powers P_e, Q_e, rotational speed ω, mechanical power P_m, rotor angle θ, electromagnetic torque T_e, and torque angle δ.

The six-phase PMSG is employed in the model **PMSG_6phN**, which are made with the purpose to test a validity of the developed model. The generator has the power of 3.38 MVA and the rated voltage of 4 kV; its parameters are taken from Ref. [14]: $L_d = 9$ mH, $L_q = 21.84$ mH, $R_s = 24.25$ mΩ, $\Psi_r = 6.73$ Wb, $f = 40$ Hz, the rotational speed is 400 r/min, $Z_p = 6$. PMSG is loaded with two loads, each of them has parameters $R_L = 8.56$ Ω, $L_L = 16.52$ mH. It is obtained in pu after simulation $T_e = P_{mex} = P_e = 0.205$; $Q_e = 0.095$; $I_{abc} = I_{xyz} = 0.24$; $I_{qs} = 0.16$; $I_{ds} = 0.18$; $P_L = 333$ kW for one winding; $U_{Lampl} = 1510$ V; $I_{Lampl} = 160$ A; $\delta = 23.16°$.

At that, the base value of the voltage is $V_b = \dfrac{\sqrt{2} \times 4000}{\sqrt{3}} = 3266$ V, of the current is $I_b = \dfrac{3.38 \times 10^6}{1.5 \times 3266} = 690$ A, and of the resistance is $R_b = 4.73$ Ω. Let's compute the theoretical values of the given quantities using relationships from Ref. [14] (*Case Study 3–4—Steady-State Analysis of Stand-Alone SG with RL Load*), where $\omega_r = 400 \times \pi/30 \times 6 = 251.3$ rad/s.

$$I_{qs} = \frac{\omega_r \Psi_r (R_s + R_L)}{(R_s + R_L)^2 + \omega_r^2 (L_L + L_d)(L_L + L_q)} = 107 \text{ A} = 0.155 \text{ (pu)} \qquad (1.119)$$

$$I_{ds} = \frac{\omega_r (L_L + L_q)}{(R_L + R_s)} I_{qs} = 120.3 \text{ A} = 0.174 \text{ (pu)} \qquad (1.120)$$

$$I_s = \sqrt{I_{qs}^2 + I_{ds}^2} = 161 \text{ A} \qquad (1.121)$$

$$V_{ds} = -R_s I_{ds} + \omega_r L_q I_{qs} = 585.2 \text{ V} \qquad (1.122)$$

$$V_{qs} = -R_s I_{qs} - \omega_r L_d I_{ds} + \omega_r \Psi_r = 1416.7 \text{ V} \qquad (1.123)$$

$$V_s = U_{Lampl} = \sqrt{V_{qs}^2 + V_{ds}^2} = 1533 \text{ V} \qquad (1.124)$$

The full power

$$P_e = P_m = 2 \times 1.5 \times Z_p \times \left[\Psi_r I_{qs} + (L_q - L_d) I_{qs} I_{ds} \right] \times \omega_r / Z_p = 668 \text{ kW}. \qquad (1.125)$$

These values obtained for the case of the absence of a magnetic coupling between two three-phase windings are quite close to the simulation results given above. Since $I_f L_{md} = \Psi_r = 6.7 \times 251.3/3266 = 0.515$ (pu), the voltage in pu is 0.469, and L_d in pu equal to $9 \times 10^{-3} \times 251.3/4.73 = 0.478$, the angle δ according to Equation (1.110) is

$$\delta = \arcsin\left[\frac{0.205 \times 0.478}{0.515 \times 0.469}\right] = 23.9°,$$

that is also very close to the value obtained by simulation, which confirms the usefulness of the developed model.

The results will differ slightly if the simulation is repeated taking into account the magnetic coupling of three-phase windings with $L_{sm} = 3$ mH and $L_{sdq} = 0.78$ mH, such as $T_e = P_{mex} = P_e = 0.186$; $Q_e = 0.087$; $I_{abc} = I_{xyz} = 0.23$; $I_{qs} = 0.147$; $I_{ds} = 0.175$; $P_L = 300$ kW for one winding; $U_{Lampl} = 1440-1480$ V; $I_{Lampl} = 150$ A; $\delta = 24.67°$, some asymmetry appears in the current curves. Thus, for accurate simulation, the coupling of the windings must be taken into account. The considered model is applied in Chapter 2 in the model **Wind_PMSG_4N**.

References

1. MathWorks, Simscape™ Electrical™, User's Guide (Specialized Power Systems). MathWorks, Natick, MA, 1998–2019.
2. Perelmuter, V. M. Electrotechnical Systems, Simulation with Simulink and SimPowerSystems. CRC Press, Boca Raton, FL, 2013.
3. Perelmuter, V. M. Renewable Energy Systems, Simulation with Simulink and SimPowerSystems. CRC Press, Boca Raton, FL, 2017.
4. Kundur, P. Power System Stability and Control. McGraw-Hill, New York, 1994.
5. Bose, B. K. Modern Power Electronics and AC Drives. Prentice Hall PTR, Upper Saddle River, NJ, 2002.
6. Hassanpoor, A., Norrga, S., Nee, H-P., and Angquist, L. Evaluation of different carrier-based PWM methods for modular multilevel converters for HVDC application. 38th Annual Conference on IEEE Industrial Electronics Society, IECON, 2012, 388–393.
7. Tu, Q., Xu, Z., and Xu, L. Reduced switching-frequency modulation and circulating current suppression for modular multilevel converters. IEEE Transactions on Power Delivery, 26(3), 2011, 2009–2017.
8. Li, Z., Wang, P., Chu, Z., Zhu, H., Luo, Y., and Li, Y. A novel inner current suppressing method for modular multilevel converters. IEEE Energy Conversion Congress and Exposition (ECCE), 2012, 4506–4512.
9. Zhao, Y., and Lipo, T. Space vector PWM control of dual three-phase induction machine using vector space decomposition. IEEE Transactions on Industry Applications, 31(5), 1995, September/October, 1100–1109.
10. Bojoi, R., Lazzari, M., Profumo, F., and Tenconi, A. Digital field-oriented control for dual three-phase induction motor drives. IEEE Transactions on Industry Applications, 39(3), 2003, May/June, 752–769.

11. Schiferl, R. F., and Ong, C. M. Six phase synchronous machine with AC and DC stator connections: Part I: Equivalent circuit representation and steady-state analysis. IEEE Transactions on Power Apparatus and Systems, 102(8), 1983, August, 2685–2693.

12. Krause, P. C., Wasynczuk, O., and Sudhoff, S. D. Analysis of Electric Machinery. IEEE Press, Piscataway, NJ, 2002.

13. Contin, A., Grava, A., Tessarolo, A., and Zocco, G. A novel modeling approach to a multi-phase, high power synchronous machine. International Symposium on Power Electronics, Electrical Drives, Automation and Motion, SPEEDAM, 2006, S19-7–S19-12.

14. Wu, B., Lang, Y., Zargari, N., and Kouro, S. Power Conversion and Control of Wind Energy Systems. John Wiley & Sons, Inc., Hoboken, NJ, 2011.

2

Wind Generator System Simulation

2.1 Fundamentals

Nowadays, the fabrication of electric energy with the help of wind force is the most popular method to obtain a clean and renewable electrical energy. According to Ref. [1], more than 52 GW of clean, emissions-free wind power was added in 2017, bringing total installations to 539 GW globally. Wind penetration levels continue to increase rapidly. In 2017, wind supplied 11.6% of the EU's power. At this, one can see dramatic price reductions for both onshore and offshore wind units. According to the forecast, the global wind generator (WG) power will reach about 840 GW by 2022.

The power of the modern WG reaches 10 MW, Ref. [2], but the WGs of much less power (dozens of kW) are utilized very extensively that are intended for supplying the remote and isolated installations.

The WG unit consists of a tower with a nacelle above, in which the main equipment is placed. The output step-up transformer is installed usually within the tower, at its base. The modern WG units have a horizontal axis of rotation with a three-blade rotor. The electric or the hydraulic drive is used in order to rotate the nacelle in the wind direction. The main WG equipment in the nacelle is schematically displayed in Figure 2.1.

In WGs of large power, the blade length can reach 70 m and more; the tower height can be more than 100 m. The blades have a rather complicated design, because the WGs of large and moderate power are equipped with devices for rotating the blades about their long axes, with the aim to change the angle of attack β and, in this way, to change the power extracted from the wind and to limit the WG power under the strong wind, exceeding the accepted value.

The rotating speed of a rotor is not more than 15–25 r/min, whereas the standard electric generators have the rated rotating speed of 1.5–2 orders greater. Therefore, the speed-up gearbox with the high gear ratio is set between the rotor and the generator, at triple stage often. Such gearboxes are bulky, have a heavy weight, and need regular maintenance. Therefore, WGs without a gearbox, with direct connection of the rotor and the generator, are developed. The special generators with the large number of pole pairs,

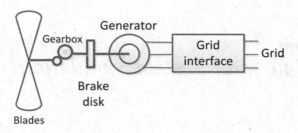

FIGURE 2.1
Main WG equipment in the nacelle.

synchronous with the wounded rotor or with the permanent magnets, are used for this purpose.

Because such generators have a large diameter, the compromised design can be utilized, in which the generator has more poles than in the usual design, but their number is not sufficient for gearless connection, so that the single-stage gearbox can be used.

Squirrel cage induction generators (SCIG), wound rotor induction generators [WRIG or doubly fed induction generators (DFIG)], synchronous generators with the excitation winding on the rotor (WRSG), and with a permanent magnet (PMSG) are employed in WG units. They connect to the power system with the help of many interfacing devices. Depending on their ability to control speed and on the type of power control, WGs are subdivided in four dominating types of wind turbines.

Type A denotes the fixed-speed wind turbine with a SCIG directly connected to the grid via a transformer (Figure 2.2a). Since the SCIG always draws the reactive power from the grid, a capacitor bank is used for reactive power compensation. A smoother grid connection is achieved by using a soft starter. Such WGs are the units with a fixed rotating speed that are defined by the grid frequency, because the slip changes not more than 1–1.5% under power alteration. As it will be seen further, in order to harvest more wind energy, the WG rotating speed must change with variation of the wind speed that is not provided in this WG; the main advantages of these WGs are their simplicity and less cost.

Type B provides limited variable speed with a variable generator rotor resistance; a wounded rotor induction generator (WRIG) is employed in these WGs. Typically, the speed range is 0–10% above synchronous speed (Figure 2.2b).

In up-to-date WGs, for optimizing the mode of operation, the semiconductor converters are employed. Two converters are usually used: the first is connected to the generator and the second is connected to the grid, with a coupling capacitor between them. The function of the first converter is to control, in an optimal way, the generator rotating speed; and the function of the second one is to transfer the fabricated energy to the power system with the control of the reactive power.

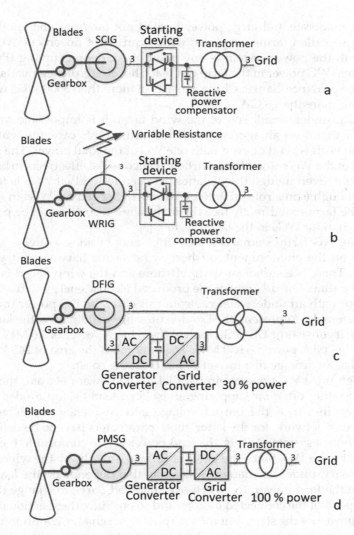

FIGURE 2.2
Main configurations of the WG. (a) Type A, (b) type B, (c) type C, and (d) type D.

Type C uses the DFIG that provides a limited variable speed wind turbine and a partial scale frequency converter (rated at approximately 30% of nominal generator power) on the rotor circuit (Figure 2.2c). The smaller frequency converter makes this concept attractive from an economical point of view.

Type D corresponds to the full variable speed wind turbine, with the generator connected to the grid through a full-scale frequency converter. The frequency converter is able to perform both the reactive power compensation and the smoother grid connection. The generator can be excited electrically for WRSG or by a permanent magnet for PMSG (Figure 2.2d).

WGs of moderate and large power located not far from one another are united in so-called farms, or parks, which consist of dozens of WGs connected with the power system at one point of common coupling (PCC), so that the total WG power in the farm can reach hundreds of megawatts. A WG farm in the province Gansu, China consists of more than 3500 WGs with the total power more than 6 GW.

Naturally, under simulation of the wind farm, it is impossible to model every WG, therefore, an aggregation is utilized; in this case, the wind park is replaced with several or even with one WG of the total power. The mutual influence of the WG rotors due to airflow (called wake effect) has to be taken into account. Even though the distance between individual WGs is taken to be high enough (some rotor diameters), the wind speed for WGs in the first rows of the farm that directly face the main wind direction can be perceptibly more than for WGs in the last row, Ref. [3].

Since the WG farms demand a large area and, besides, exercise negative influence on the environment, offshore wind farms have been developed and built. There is a sufficient space offshore and the wind speed is higher most of the time. The delivery of the produced electric energy to the shore is carried out with an underwater cable; in this case, the DC power transmission is preferable. At present, the London offshore wind park is the largest in the world. It consists of 175 WGs; each of them has a power of 3.6 MW, so that the total WG park power is 630 MW. The park takes the area of 245 km² and is located at an average distance of 20 km from the coast.

Although most WGs operate in parallel to the power network, the WG is employed rather often for supplying an isolated load (island mode). Unlike the former, in which the output voltage and frequency are determined by the power network, for the latter, these parameters have to be defined by the WG control system. At this, the load power has to conform to the power produced by the WG. Since the latter is defined, mainly, by the wind speed, it is necessary, under its change, in the way that depends on the load type, to change the load power: to switch over the load parts, to change the load and the speed of the connected motors, and so on. Since these actions change the load power by the steps, but the WG power can change continuously and exceed the load power, measures have to be taken, to limit the WG rotating speed. If the WG is equipped with a device for blade position change, it begins to operate, limiting the WG power. This device is usually absent in a low power WG; in such a case, the dummy (ballast) load must be used.

The wind turbine power may be written as

$$P_m = 0.5\rho A V_w^3 C_p(\lambda, \beta) = K_p V_w^3 C_p(\lambda, \beta), \qquad (2.1)$$

where ρ is the air density that is equal to the average of 1.2 kg/m³, A is the sweep area of the turbine rotor blades that is equal to πR^2, R is the blade length, V_w is the wind speed, C_p is the power coefficient that depends on the blade angle position β and on the tip speed ratio λ, $\lambda = R\omega_r/V_w$, and ω_r is the rotor rotating speed ($2 < \lambda < 13$ usually).

The dependence of C_p on λ has a maximum under λ_m; thus, to harvest maximum wind power, the WG rotation speed has to vary with the wind speed. The value $C_{pmax} = 0.593$ theoretically and reaches the values 0.35–0.5 in the modern WG. The different analytical expressions for C_p (λ) exist. In SimPowerSystems, the following relationship is accepted

$$C_p(\lambda,\beta) = c_1\left(\frac{c_2}{z} - c_3\beta - c_4\right)e^{-c_5/z} + c_6\lambda, \tag{2.2}$$

$$\frac{1}{z} = \frac{1}{\lambda + 0.08\beta} - \frac{0.035}{1+\beta^3} \tag{2.3}$$

$$c_1 = 0.5176, \quad c_2 = 116, \quad c_3 = 0.4, \quad c_4 = 5, \quad c_5 = 21, \quad c_6 = 0.0068.$$

The dependencies of C_p on λ calculated by Equations (2.2) and (2.3) under the various β values are given in Figure 2.3. Value $C_{pmax} = 0.48$ under $\lambda_m = 8.1$.

The turbine torque is defined by the following expression

$$T_m = K_p R V_w^2 C_p(\lambda,\beta)/\lambda = K_m V_w^2 C_m(\lambda,\beta), \tag{2.4}$$

where $K_m = K_p R$. The plot C_m (λ, β) = C_p (λ, β)/λ for C_p by Equation (2.2) is depicted in Figure 2.4.

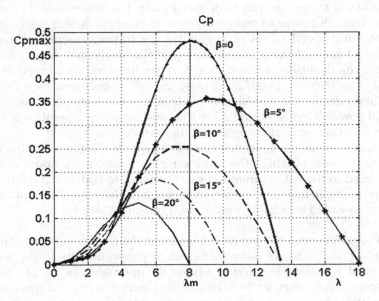

FIGURE 2.3
Dependencies of C_p on λ and β.

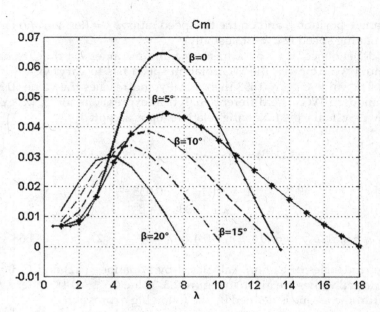

FIGURE 2.4
Dependencies of C_m on λ and β for expression (2.2).

Thereby, under each wind speed, the rotor speed exists when the turbine power is maximal (Figure 2.5). At some small wind speed (cut-in speed), the turbine starts to operate and to deliver power. The blade should be able to capture enough power to compensate for the turbine power losses. At the definite wind speed (12–14 m/s usually,) the turbine reaches its nominal power, and under further increasing wind speed, the power is limited by the action on the blade angle position β. When the wind speed reaches some critical value (cut-out speed; 20–25 m/s), the WG stops, turns away from the wind, and is braked. It should be noted that in practice, a transition from the optimal power curve in Figure 2.5 to the constant power operation at higher speeds is not carried out abruptly, but more smoothly.

The main task of the control system is to provide the rotor rotating speed equal to the one, at which the turbine power is maximal, under different wind speed in the cut-in range—nominal, that is, to realize the relationship corresponding to the curve shown in Figure 2.5. There are different ways to perform so-called maximum power point tracking (MPPT).

1. The wind speed measurement is used in the first group of methods. It is possible to set the rotor speed ω_r proportionally to the wind speed V_w, taking in mind to achieve the optimal value λ_m. The rotor speed is controlled with the help of the power inverter and speed controller. The other opportunities are to set the turbine power according to relationship (2.1) or its torque according to Equation (2.4) with

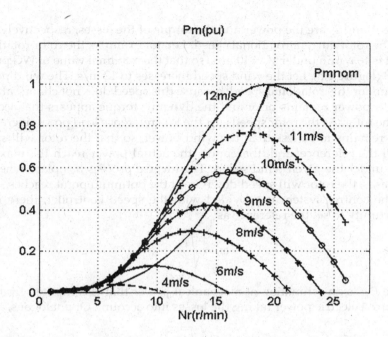

FIGURE 2.5
An example of dependence of WG power on the wind speed and the turbine rotation speed.

$C_p = C_{pmax}$. In practice, instead of theoretical relationships, the optimal power versus wind speed curves for a given wind turbine provided by the manufacturer are employed.

2. The accurate and stable measurement of the wind speed is difficult to provide, allowing for fluctuating nature the wind. Therefore, the other methods, which do not require the V_w measurement, are preferred. Using the relationship for λ and Equation (2.1), the expression for the maximal power can be written as

$$P_{mmax} = (K_p C_{pmax} R^3 / \lambda_m^3)\omega_r^3 = K_{p1}\omega_r^3. \tag{2.5}$$

Analogous, the expression for the optimal torque can be written as

$$T_{mopt} = (K_p C_{pmax} R^3 / \lambda_m^3)\omega_r^2 = K_{p1}\omega_r^2. \tag{2.6}$$

There are two ways to use these relationships. If there is the possibility to control WG electromagnetic power or torque, then the reference values of the power or torque can be given as

$$P_e^* = K_{p1}\omega_r^3 - P_l \tag{2.7}$$

or

$$T_e^* = K_{p1}\omega_r^2 - T_l, \tag{2.8}$$

where P_l and T_l are the power and the torque of the losses, respectively, and ω_r is the rotor current rotational speed. Let, for example, the rotor rotational speed is 16 r/min under $V_w = 10$ m/s, so that the maximal value of WG power is 0.58 (Figure 2.5). Let the wind speed increases to 12 m/s. The wind power is 0.9 under this rotational speed; because this speed does not change at first, the WG power remains prior and the dynamic torque appears that accelerates the rotor. As the rotor speeds up, the WG reference load power increases, but it remains to be less than the wind power, so that the rotor will speed up till the reference, and, thereupon, the actual power reach the maximal value under the given wind speed. The inverse process has place when V_w decreases, the rotor will speed down until the optimum point reaches.

If the control system is provided with the speed controller, there is an opportunity to set the reference as

$$\omega_r^* = \sqrt[3]{\frac{P_{me}}{K_{p1}}},\tag{2.9}$$

where P_{me} is an estimation of mechanical power that is computed from the measured electric power taking the losses into account. Or analogous,

$$\omega_r^* = \sqrt{\frac{T_{me}}{K_{p1}}}.\tag{2.10}$$

An approximation of relationship (2.9) with the second-order polynomial is used in some WGs developed by General Electric Company (GE) and is utilized in the demonstration models **Wind Farm DFIG Detailed Model (power_wind_dfig_det)** and **power_wind_type_4_det**.

Utilization of the given MPPT methods is displayed with a number of examples considered in Ref. [4].

The model **Wind Turbine** is included in SimPowerSystems; their inputs are the wind speed (m/s), the WG rotating speed in pu, the angle of the blade deviation from the optimal position (pitch, grad), and the output is the torque in pu, referring to the base quantities of the driven generator (Figure 2.6). The

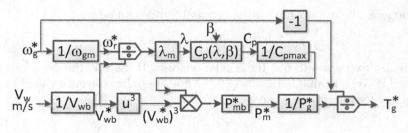

FIGURE 2.6
Block diagram of the turbine model.

moment of inertia is not taken into account and has to be added to the generator moment of inertia. It is specified in the wind turbine dialog box: the turbine nominal power, P_{mnom}; the nominal generator power, P_{gnom}; the wind base speed V_{wb} (m/s) (as such the mean value of the expected wind speed can be taken, or the prevailing wind speed in this region, or the wind speed, when the turbine nominal power is reached); the maximum power under V_{wb} in pu, P_{mb}^*; and the WG rotating speed, at which the maximum power for V_{wb} is obtained, in pu relative to the generator base speed ω_{gm}. These parameters are used in the following way. Because

$$P_{mb} = P_{mb}^* P_{mnom} = K_p V_{wb}^3 C_{p\max}, \qquad (2.11)$$

the relative turbine power P_m^* for the some wind speed V_w may be written as

$$P_m^* = P_{mb}^* C_p V_{wb}^{*3} / C_{p\max}, \qquad (2.12)$$

where $V_{wb}^* = V_w / V_{wb}$.
The quantity λ in the formula for C_p is

$$\lambda = \frac{\lambda_m \omega_r^*}{V_{wb}^*} \qquad (2.13)$$

$$\omega_r^* = \frac{\omega_g^*}{\omega_{gm}}, \omega_g^* = \omega_r / \omega_{gnom}.$$

The output quantity is the mechanical torque in pu of the nominal generator torque:

$$T_g^* = -\frac{P_m^*}{P_g^* \omega_g^*}, \qquad (2.14)$$

where $P_g^* = P_{gnom} / P_{mnom}$.
It is seen that the sign of the torque is negative, which is suitable when the induction generator (IG) is employed. When SG is utilized, the power, not the torque, comes to its input P_m; the quantity $T_g^*(-\omega_g^*)$ has to be sent at this input. When the PMSG is used, the torque in SI unit has to come to its input T_m: $T_g^* P_{gnom} / \omega_{gnom}$.
The wind turbine model is defined in SimPowerSystems with its static characteristics; the processes caused, for example, by elasticity are not taken into account. Besides, it is supposed that the wind force affects the blades uniformly, regardless of their position; in fact, it is not a case, these factors will be considered in the next section.

2.2 Modeling Wind Shear and Tower Shadow Effect

Even under the constant wind speed in the free air, the periodic processes exist in the WG that are excited by instability of the local wind speed in the different points of the WG set; they are caused mainly by the wind shear and tower shadow. The term wind shear is used to describe the variation of wind speed with height, while the term tower shadow describes the redirection of wind due to the tower structure, Ref. [5].

It is proved in the published articles that an impact of the first factor is much less than the second one, so that the latter will be considered. The main design parameters that effect tower shadow are as follows: R is the blade length, a is the tower radius, and x is the distance from the blade origin to the tower midline. The tower effect is maximal when one of the blades directs straight down, that is, the influence of this factor has a maximum three times for one blade rev. It is shown in Ref. [5] that the tower shadow torque relative to the turbine torque (calculated without taking this effect into account) can be computed as

$$\Delta T = \frac{2}{3R^2} \sum_{i=1}^{3} \left[\frac{a^2}{\sin^2 \theta_i} \ln \left(1 + \frac{R^2 \sin^2 \theta_i}{x^2} \right) - \frac{2a^2 R^2}{x^2 + R^2 \sin^2 \theta_i} \right]. \tag{2.15}$$

Here θ_i is the angular position of i-blade:

$$\theta_1 = \theta = \int \omega_r \, dt, \qquad \theta_2 = \theta - \frac{2\pi}{3}, \qquad \theta_3 = \theta + \frac{2\pi}{3}$$

where ω_r is the rotor angular velocity. The corresponding term is not null, if only $\pi/2 < \theta i < 3\pi/2$.

To simulate this effect, the subsystem **Tower Shadow** in the model **Tower** is made. It is supposed that rotor rotational speed is given in pu relative to the generator base rotational speed that is equal to $2\pi f/Z_p$, where f is the generator voltage nominal frequency and Z_p is the pole pairs number.

The WG with the DFIG with a power of 1.6 MVA under a frequency of 50 Hz and $Z_p = 2$ is considered. The turbine with a nominal power of 1.5 MW produces 0.73 of nominal power at the wind base speed of 10 m/s; the base rotational speeds of the turbine and the generator are the same. Then the nominal wind speed is $10/\sqrt[3]{0.73} = 11.1$ m/s, and generator rotational speed is

$$\omega_g = \frac{2 \times \pi \times 50 \times 1.11}{2} = 174.27 \text{ rad/s}.$$

If to take $R = 45$ m, then the rotor rotational speed is $\omega_r = 8.1 \times 11.1/45 = 2$ rad/s under $\lambda_m = 8.1$; therefore, the gear reduction rate is $i_r = 174.27/2 = 87.135$. The other parameters are seen in the dialog box of the subsystem **Tower Shadow**. The normalized torque due to tower shadow is depicted in Figure 2.7 under

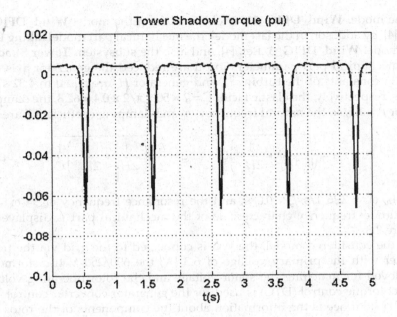

FIGURE 2.7
Plot of normalized torque due to tower shadow.

the nominal wind speed. It is seen that the torque frequency is about 1 Hz and the amplitude is about 7%, but the integral action of tower shadow is not large.

Nevertheless, it can have a bad effect on WG function, having in mind the elastic nature of the mechanical part. The latter, for study of elastic vibrations, can be reduced to a two-mass model, Ref. [6]. It is supposed that the torque T_{sh} transferred through the shaft may be defined as

$$T_{sh} = C \int (\omega_1 - \omega_2) dt + B(\omega_1 - \omega_2), \qquad (2.16)$$

where ω_1 is the rotational speed of the shaft input point, ω_2 is the same for the shaft output point, and C and B are the factors. These speeds may be found as

$$\frac{d\omega_1}{dt} = \frac{T_e - T_{sh}}{J_1} \qquad (2.17)$$

$$\frac{d\omega_2}{dt} = \frac{T_{sh} - T_l}{J_2}, \qquad (2.18)$$

where T_e is the torque applied to the shaft input (the torque of the prime mover), J_1 is the moment of inertia of the prime mover, T_l is the torque of the resistance at the output shaft end, and J_2 is the moment of inertia of the rotating mass connected to the shaft output end.

The model **Wind_DFIG_1N** is obtained from the model **Wind_DFIG_1**, Ref. [4], by addition in the last model, the mechanical part model taking from the model **Wind_DFIG_3**, Ref. [4], and also the subsystem **Tower Shadow**; the mechanical part parameters in pu are reduced the generator axis. The inertia constants of the turbine H_t and generator H_g are equal to 4.32 s and 0.69 s, respectively, the elastic factor $C = 2 \times 50 \times \pi/2 \times 0.4 = 62.8$, the damping factor $B = 1.5$, so the natural frequency ω_0 and damping coefficient ζ are

$$\omega_0 = \sqrt{\left(\frac{1}{2H_t} + \frac{1}{2H_g}\right) C}, \qquad \xi = \frac{B}{2}\sqrt{\left(\frac{1}{2H_t} + \frac{1}{2H_g}\right)\frac{1}{C}} \qquad (2.19)$$

that is, $\omega_0 = 7.26$ 1/s, $\zeta = 0.086$, and the resonance frequency $\omega_{res} \approx \omega_0$. The amplitude–frequency characteristic of the mechanical part is displayed in Figure 2.8.

In the considered model, the WG is connected to the grid via the transformer with the primary voltage of 6.3 kV; the 690/250 V transformer is employed for matching the stator voltage and the rotor operating voltage. Direct torque control (DTC) is used for the generator converter control. The DTC system needs the information about the components of the rotor flux linkage space vector in the stationary reference frame and about the DFIG torque. These data can be obtained with the help of the so-called voltage and

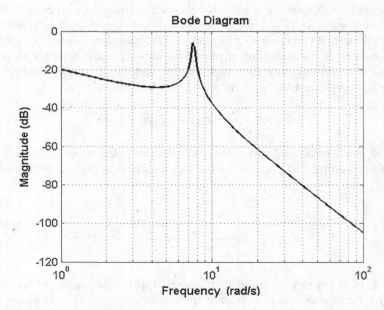

FIGURE 2.8
Amplitude–frequency characteristic of the mechanical part of the model **Wind_DFIG_1N**.

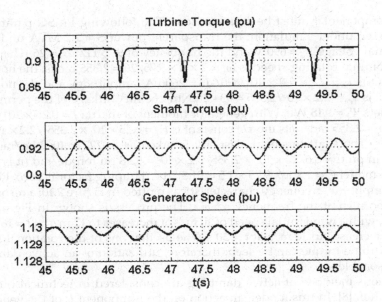

FIGURE 2.9
Turbine torque, generator shaft torque, and its speed vibrations owing to tower shadow.

current models. Their description is given in Ref. [4]. With the tumbler in the subsystem **Control_VSI-Ge** in the top position, the voltage model is used; in the low position, the current model is used.

The turbine torque, generator shaft torque, and its speed vibrations are shown in Figure 2.9. It is seen that, because of the large inertia of the rotating details, the torque and speed alternations are not essential. Such a situation is typical for the modern WG, but the condition can change with the development of the new, lighter blade materials, Ref. [7].

2.3 Active Damping Simulation

The elastic properties of the WG mechanical part have already been taken into account in the previous model, which be considered as a two-mass system. The elastic nature manifests itself, especially in the WG with direct- (without gear) connection generator and turbine, Ref. [8]. Such a system is simulated in the model **Wind_PMSG_1N** that is the revision of the model **Wind_PMSG_1** from Ref. [4], in which the subsystem **Drive Train** is added in the model WG. The surface permanent magnet synchronous generator (SPMSG) is utilized; therefore, the wanted torque defines the generator current component I_q and the component I_d must be equal to zero. The following PMSG parameters are taken under simulation: the transparent power $S = 2.2$ MVA under the nominal voltage $V_n = 690$ V and the frequency of 9.75 Hz, $Z_p = 26$. Therefore, the nominal rotating speed is $\omega_n = 2 \times \pi \times 9.75/26 = 2.355$ rad/s, the nominal current is $I_n = \sqrt{2} \times 2.2 \times 10^6/690/\sqrt{3} = 2606$ A (amplitude), and the nominal torque is $T_{en} = 2.2 \times 10^6/2.355 = 934.2$ kNm, the flux linkage of the permanent magnets $\Psi_r = 9.18$ Wb. With generator moment of inertia $J = 0.25 \times 10^6$ kg m² and $\omega_n = 2.355$ rad/s, its inertia constant is $H_g = 0.25 \times 10^6 \times 2.355^2/2.2 \times 10^6/2 = 0.315$ s. The shaft torsional elasticity is taken as 8.6727×10^7 Nm/rad that gives in pu $C = 8.6727 \times 10^7 \times 2.355/2.2 \times 10^6 = 92.8$ pu torque/rad (it is taken in the model $C = 2 \times \pi \times 9.75 \times 1.5 = 91.8$), the damping factor is zero, Ref. [8]. The amplitude–frequency characteristic is plotted in Figure 2.10; a resonance may be seen at the frequency ≈ 2 Hz. The process is shown in Figure 2.11 when, with a wind initial speed of 10 m/s, it increases to 12 m/s at $t = 16$ s and then decreases to 10 m/s at $t = 30$ s. The oscillations of the rotational speed and turbine torque can be seen; therefore, utilization of an active damping is reasonable.

Various methods of active damping are considered in technical publications, Ref. [8]. In this model, an action on the component I_q of the generator current (it means, on its torque) by the rotational speed signal is used, direct or after the high-frequency filter having the band-pass of 2 Hz, in addition

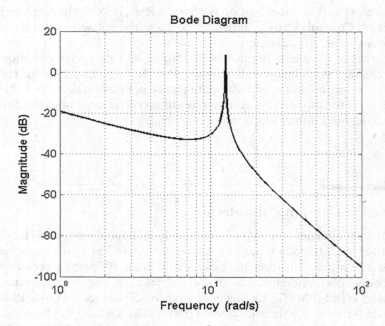

FIGURE 2.10
Amplitude–frequency characteristic for the model **Wind_PMSG_1 N**.

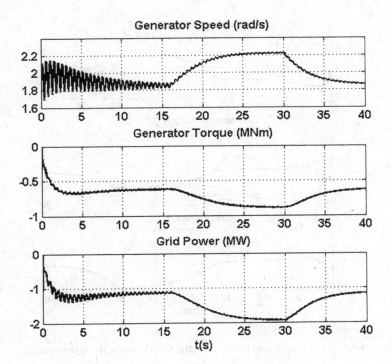

FIGURE 2.11
Processes in the model **Wind_PMSG_1N** without active damping.

(with the sign minus) to the speed controller output (Figure 2.12). Variations of the speed, torque, and power with employment of the first option are shown in Figure 2.13. One may see that the oscillations are put down almost completely. But the additional signal intensity is rather large that demands an increase in the operative range of the speed controller and can tell on system dynamic characteristics. For the second option, the oscillations are somewhat higher (Figure 2.14), but they are much lower than without active damping.

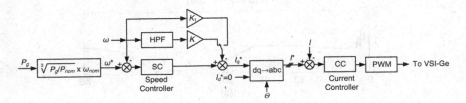

FIGURE 2.12
Block diagram of the active damping.

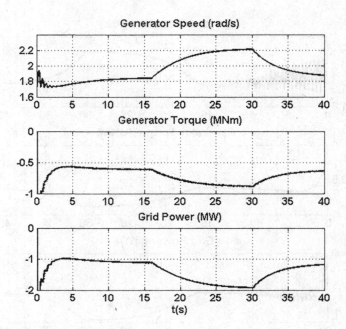

FIGURE 2.13
Variations of the speed, torque, and power with the first option of the active damping.

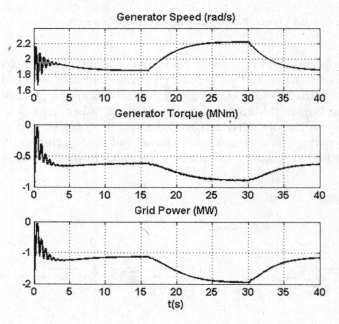

FIGURE 2.14
Variations of the speed, torque, and power with the second option of the active damping.

2.4 System Inertia Support

The main parameter that defines stability of the electrical network is the frequency deviations, which arise under the abrupt discrepancy between generated and utilized powers; at that, the voltage changes are of lesser significance because they can be parried well and fast by the generator excitation control systems. The synchronous generators (SGs), with their inertia masses and the primary mover control systems (turbine-governor systems), play a primary role in the frequency change dynamic. Insertion of WGs in the electrical power system decreases its general inertia. Variable speed wind turbines are equipped with the power electronic converters, which are intended for power transmitting to the grid. Because of the decoupled control between mechanical and electrical systems, WGs do not response to the system frequency changes, which results in an increase in the system frequency change under load variations. The importance of this factor arises with the growth of penetration of WGs into the power system; therefore, the demands raise and the intensive researches are carried out to increase WG possibilities to contribute to frequency change decrease under load change. Such WG ability is called a frequency support. Of course, one can speak about such support only in the case of the deep penetration of WGs in the power system that is defined as a ratio of the total WG power under nominal (or close) work conditions to the electrical system power. Such a situation has place often for offshore wind farms.

Evidently, to have an ability to carry out the frequency support, WGs must have some energy reserve. It can be an energy saved in the capacitors of the voltage source inverters (VSIs) and a kinetic energy of the WG rotors, Ref. [9].

However, the energy of the capacitors is not large with their reasonable size, and the utilization of the kinetic energy results in deviation of the WG power from the optimal values; besides, the WG rotational speed and, consequently, its kinetic energy decreases with the lowering wind speed, so the possibility of the frequency support reduces too. Such a system will be considered further.

A surer method is availability of the additional power, for instance, of the batteries, which are set in the DC link of the WG inverters. Such systems are implemented in practice, Refs. [10, 11].

The batteries with the power of 200 kW are set in these WGs having the power of 2 MW. Such a system with PMSG is simulated in the model **SG3b**. The electrical grid is modeled with SG having the power of 125 MVA and the step-up transformer of 13.8/230 kV. The control systems of the individual generators forming power system can be diverse; for the simplicity, the control systems for a diesel generator is accepted (Chapter 6). The grid supplies the active-inductive load with a power of 100 MW that can be increased to 130 MW at $t = t_1$. The WG park model consists of two aggregated

WG models connected in parallel. These models have the powers of $2N$ and $2N_1$ MW each at the wind speed of 12 m/s. It is taken $N = N_1 = 30$. The investigation is carried out at the wind speed of 10 m/s, so the WG park power is $2 \times 60 \times (10/12)^3 = 69.4$ MW, it connects to the grid via the step-up transformer 35/230 kV. The batteries via boost/ buck converters (Figure 1.5) connect to WG DC links, whose voltage is taken as 1200 V. The battery voltage is 400 V, the rated current 400 A (i.e., 400 N and 400 N_1 for parallel models, respectively). When the frequency drops to 49.6 Hz, the battery begins to discharge, the control pulses, which are fabricated with the help of the discharge current controller, come to the transistor T2. Two variants are provided to specify discharge current, which are selected by the tumbler **Manual Switch**: the grid frequency controller with the reference of 49.8 Hz (the first option) or the direct current assignation as $400 \times N$ (A) (the second option).

When the frequency arises to 49.9 Hz the discharge stops; the battery comes in the charge mode, if its charge is less as 99%, with the current $100 \times N$ (A). The pulses come to the transistor T1.

A somewhat artificial way to organize the parallel operation of the SG and WG park is utilized in the model, because in the actual conditions, the individual WGs are connected to the network not simultaneously, as in the aggregative models, but consecutively. First, SG connects to the load 100 MW, and WG connects to the temporary load 65 MW; the WG output current aligns with the SG output voltage. At $t = 4$ s, WG is connected in parallel with SG supplying the load 100 MW jointly, the temporary load disconnects. At $t = 6$ s, WG output current begins to align with its output voltage as usually; it is supposed that the transformer winding connections are made in the way to provide the necessary phasing. It is supposed that steady-state conditions reach to $t = 10$ s; the additional load 30 MW is connected at this instant. The transients are shown in Figure 2.15. It is seen that, with the second variant for control of the battery discharge current (the definite current designation), the frequency nadir halves. The WG park power increases by about 10 MW for short time, which results in a decrease in nadir.

There are also the methods for frequency support without external sources, with employment of the WG intrinsic energy, in particular, the energy that is saved in the capacitors in DC link VSIs and WG rotor kinetic energy, Ref. [9]. Thereto, when the grid frequency decreases, the capacitor voltage has to be diminished, and also the rotor rotational speed, so that the released electrostatic energy of the capacitors and the kinetic energy of the rotating rotors were directed to the grid. Such a system is simulated in the model **SG4**, which is the previous model without the battery. Slowdown is realized by a decrease in generator power reference. Unlike the system described in the reference publication, in which the frequency continuous-time controllers are used for a change of the capacitor voltage and the rotor rotational speed,

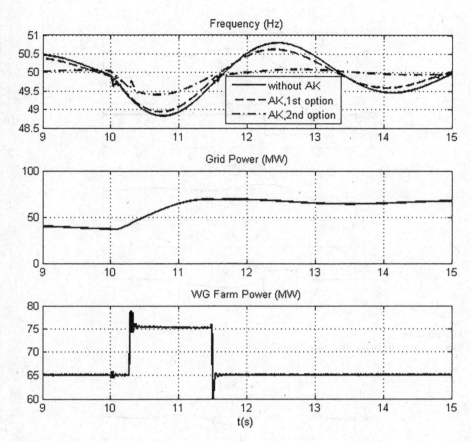

FIGURE 2.15
Frequency nadir in the system with battery.

relay controllers are utilized in the modeled system; these relays vary in step the reference values of the capacitor voltage and WG power when frequency deviates by the definite value. When the frequency drops to 49.6 Hz, the DC link voltage and the reference power decrease by 15%, the nominal mode recovers when the frequency increases to 49.9 Hz. Similarly, the DC link voltage and the reference power increase by 10% with the frequency rise to 50.4 Hz, the nominal mode recovers when the frequency decreases to 50.1 Hz.

The process under load increment (like in the previous model) is shown in Figure 2.16. One may see that the frequency nadir is much less than without frequency support (Figure 2.15): 49.3 Hz instead of 48.8 Hz. The process under load decrement is shown in Figure 2.17.

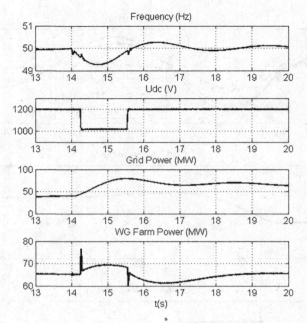

FIGURE 2.16
Frequency nadir in the system without battery.

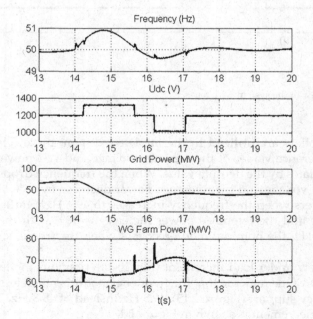

FIGURE 2.17
Frequency deviation in the system without battery under load decrement.

2.5 Synchronverter Simulation

In the models considered previously, the synchronization of the grid inverter VSI-Gr with the grid is carried out with the help of Phase Lock Loop (PLL), whose practical realization causes some difficulties, Ref. [12]. The possibilities to use another structure for VSI-Gr control that dispenses with PLL are investigated in a number of articles. Because the behavior of such an inverter mimics the behavior of SG connected to the network, this inverter received a name synchronverter (SV), Ref. [13].

Like a SG, the control system receives two control signals: a mechanical torque T_m and excitation current that is accepted as a designed value, neglecting its fluctuations, owing to the supply source and rotor oscillations; therefore, the rotor field linkage Ψ_r may be taken as given value. Then the SV is governed with the equations:

Virtual motion equations (swing equations)

$$J\frac{d\omega_s}{dt} = T_m - T_e - D_e(\omega_s - \omega_g) \qquad (2.20)$$

$$\frac{d\theta}{dt} = \omega_s. \qquad (2.21)$$

Here J is the SV virtual moment of inertia, ω_s is its frequency, ω_g is the grid frequency, D_e is the damping factor, θ is the angular position of the virtual rotor that determines the direction for projections of components of the system space vectors, T_e is the virtual SG electromagnetic torque that, for the round rotor and $Z_p = 1$, may be found as (see 1.34)

$$T_e = 1.5\Psi_r i_q = \Psi_r \langle i, \overline{\sin\theta} \rangle, \qquad (2.22)$$

where i_q is the component of the stator current vector $\mathbf{i}$, the vector product is in brackets,

$$\overline{\sin\theta} = \begin{bmatrix} \sin\theta \\ \sin\left(\theta - \dfrac{2\pi}{3}\right) \\ \sin\left(\theta + \dfrac{2\pi}{3}\right) \end{bmatrix}. \qquad (2.23)$$

The inverter three-phase internal EMF is

$$E = \Psi_r \omega_s \overline{\sin\theta}. \qquad (2.24)$$

Because the inverter internal voltage is

$$E_s = \frac{m}{2}U_{dc},\qquad(2.25)$$

where E_s is the amplitude of the phase voltage and m is the amplitude of the three-phase modulation signal (modulation index), then

$$m = \frac{2E}{U_{dc}}.\qquad(2.26)$$

The reference value of Ψ_r is adjusted by the output of the controller of the reactive power, which may be computed as

$$Q = -\Psi_r\omega_s\left\langle i, \overrightarrow{\cos\theta}\right\rangle = -1.5\Psi_r\omega_s i_d\qquad(2.27)$$

$$\overrightarrow{\cos\theta} = \begin{bmatrix} \cos\theta \\ \cos\left(\theta - \dfrac{2\pi}{3}\right) \\ \cos\left(\theta + \dfrac{2\pi}{3}\right) \end{bmatrix}\qquad(2.28)$$

The described system is realized in the model **Synchrover1**. DC voltage $U_{dc} =$ 10 kV, the inductance at the inverter output is 8 mH, the grid voltage is 6 kV. The virtual SG power is taken as 2.5 MW under the frequency of 50 Hz, therefore, the rated torque is $T_{mnom} = 2500/(2 \times 50 \times \pi) = 7.96$ kNm. Equations (2.20) and (2.21) are solved in the subsystem **Mech**. It is taken $J = 400$ kg m² that gives the value of the inertia constant $H = 400 \times 314^2/2.5 \times 10^6/2 = 7.9$ s, the damping factor $D_e = 8000$, the flux linkage is $\Psi_r = \sqrt{2} \times 6000/(\sqrt{3} \times 314) = 15.57$. The reactive power controller that affects the flux linkage is integral controller with limitation ±6, the reactive power reference is zero.

The special procedure is utilized for the initial connection of the SV to the grid (Figure 2.18). The additional integrator is used to provide equality of the frequencies of the grid and SV. In place of the absent current i, the virtual current is used

$$i_{er} = \frac{V_i - V_g}{pL + R},\qquad(2.29)$$

where V_i is the output inverter voltage (behind filter), V_g is the grid voltage, and R and L are the virtual parameters that are close to the parameters of the grid proper segment. The Ψ_r value is adjusted by the output of the V_i voltage

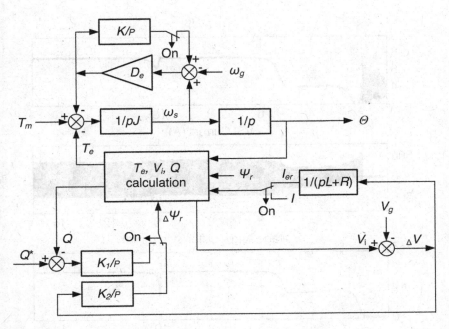

FIGURE 2.18
Block diagram of the synchronverter.

controller, whose amplitude must be equal to the grid voltage amplitude. When the difference in frequency between the grid and SV, and also the difference in their voltages decrease to the small values, the contactor closes that connects SV and the grid.

The processes of SV operation are shown in Figure 2.19. The contactor closes at $t \approx 2$ s; at that, the current surge some more than the steady-state value. The torque reference increases from 4 kN to 7.5 kN at $t = 5$ s. The power delivered in the grid increases from 1.2 MW to 2.3 MW; power rise owes increasing difference in the angles between the space vectors of the grid and inverter voltages from 7.5° to 14°. The phase of the grid voltage steps by 5° at $t \approx 10$ s; but inverter voltage phase changes by the same value, so its power and torque do not vary. The inverter reactive power is close to zero.

The above investigated SV is employed for optimization of WG having power $P_w = 2.5$ MW at the wind speed of 12 m/s in model **Synchrover2**. The generator with power $P_{gen} = 3$ MVA has six pole pairs and rated rotational frequency $\omega_{nom} = 60.3$ rad/s. The DC/DC boost converter is set at the generator output; the converter control system keeps its output voltage (DC link voltage). It is accepted $U_{dc} = 10$ kV. SV is taken from the previous model with alternations in the circuits for connection to the grid 120 kV

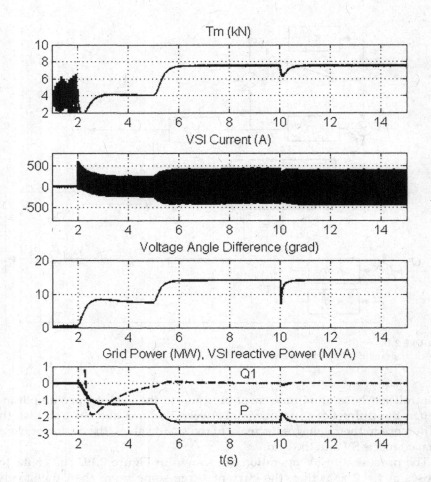

FIGURE 2.19
SV operation.

via the step-up transformer 5/120 kV. The torque reference is $T_m = P^*/\omega_s$, where

$$P^* = \left(\frac{\omega}{\omega_{nom}}\right)^3 P_w, \tag{2.30}$$

ω is the WG current rotational speed; MPPT is obtained with this way.

The process is displayed in Figure 2.20 when, with the initial wind speed of 10 m/s, it increases to 12 m/s at $t = 6$ s. It is seen that optimal point tracking is provided (taking losses into account), DC link voltage is equal to the reference value.

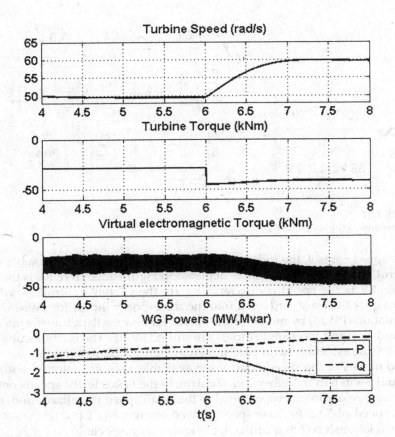

FIGURE 2.20
SV operation in WG set.

2.6 Wind Turbines with Multiphase Generators

Some WG models are given below, in which the multiphase generators are utilized, which were described in Chapter 1.

The six-phase IG with the squirrel-cage rotor is used in the model **WIND_IG_6N**; IG has a power of 1200 kVA, each winging voltage is 690 V. Two back-to-back connected VSIs are set in the circuits of the each winding (Figure 2.21). The converters C1 and C2 control the WG rotating speed; the converters C3 and C4 maintain the voltages across the capacitors; the reference value is 2 kV. They are connected in parallel via reactors and are connected with the 35 kV/1200 V step-up transformer, which delivers the fabricated electric energy into the network having the short-circuit power of 150 MVA.

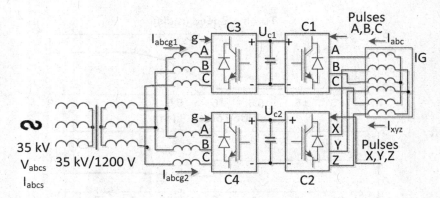

FIGURE 2.21
Main circuits of the six-phase IG.

For speed control, the indirect vector control is realized in the subsystem **Control**, whose block diagram is depicted in Figure 2.22. Here RL is the rate limiter; SR is the speed controller; CRs are the current controllers, whose output quantities are converted into the three-phase inputs for pulse-width modulation (PWM) by means of the blocks for reverse Park transformations Tp3 and Tp4; and their angle inputs are shifted by 30°. The relationships that describe the system function are shown in Figure 2.22, Ref. [14].

Two modes of the speed control are available; they are selected with the **Manual Switch** in the subsystem **Control**; in the first one, the speed controller, whose reference is proportional to the wind speed is utilized: under the wind speed of 10 m/s (base speed), speed reference is 1 pu. In the second mode, relationship (2.8) is utilized, the losses are neglected.

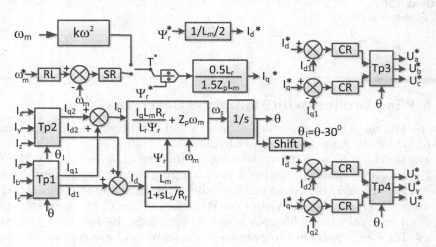

FIGURE 2.22
Block diagram of the IG indirect control.

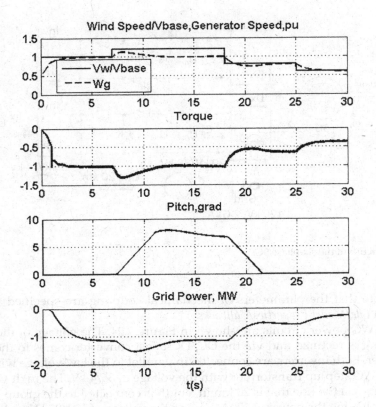

FIGURE 2.23
Processes in the model **WIND_IG_6N**.

In the control system of converters C3 and C4, the voltage across each capacitor is controlled separately, but the block PWM with the triangle waveform with the frequency F_c carries out the shift of the waveforms for both controllers by $0.5/F_c$; because of that the content of the harmonics in the network decreases.

The processes in the system for the second control mode are displayed in Figure 2.23. It is seen that MPPT is provided; the pitch control limits efficiently the WG power under large wind speed. The investigation with the help of **Powergui** option *FFT Analysis* proves that content of the high harmonics (total harmonic distortion [THD]) in the generator winding currents and in the grid currents is less than 5%.

The six-phase SG is employed in the model **Wind_SG_2aN**. The wind turbine has the power $P_n = 3.6$ MW at the base wind speed $V_w = 12$ m/s. SG power is 4 MVA, it has three pole pairs, the voltage of the each winding is 2400 V. The turbine-generator mechanical system is considered as two-mass one, the turbine and generator inertia constants are 0.9 s and 0.45 s, respectively, the factors C and B are equal to $C = 2\pi f K_{sh}/Z_p = 100 \times \pi \times 1.1/3 = 115$ pu, $B = 1.5$ pu.

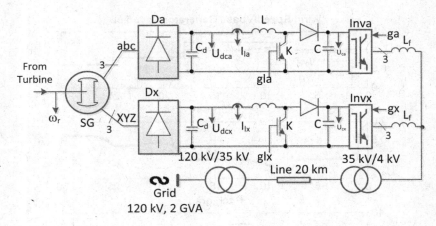

FIGURE 2.24
Main circuits of the six-phase SG.

We note that the parameters of mechanical coupling are specified in the option *File/Model Properties/Callbacks*.

The WG power circuits are shown in Figure 2.24. The voltage of the each winding is rectified and via the DC/DC boost converter comes to the grid inverter. Both inverters are connected in parallel to the low-voltage winding of the WG step-up transformer with the voltage of 35/4 kV. The high-voltage winding via the line that is 20 km in length is connected to the group transformer having the power of 100 MVA and the voltage of 120/35 kV and, afterward, to the network with the power of 1000 MVA.

MPPT is provided with a proper control of the boost converter currents (in the converter inductances). Their references are

$$I_{la(x)}^* = (\omega_m^2 - a - b\omega_m)\omega_m \frac{P_n}{2U_{dca(x)}}, \tag{2.31}$$

where ω_m is the SG rotating speed in pu, a, b are the loss factors, and $U_{dca(x)}$ are the rectified voltages. Because the maximal power at base wind speed in pu and the rotating speed in pu are accepted equal to 1, the WG is loaded with the power that is maximal under the actual rotational speed; by this way, MPPT is provided, see formula (2.7).

The grid inverters keep the voltages at the boost converters outputs at the levels of 6 kV and in addition control the network reactive power. In order to decrease the content of the higher harmonics in the network current, the triangular waveforms of both PWM units are shifted mutually by one-half of their period.

The excitation control system produces the stator flux linkage nominal value F_{smod} when the rotating speed is not more than the nominal value, and the flux linkage lowering when speed excess appears. The calculation of F_{smod} is carried out by relationships, Ref. [15]

$$\Psi_{s\alpha} = \int (U_{s\alpha} + \lambda U_{s\beta} - \lambda\omega_s\Psi_{s\alpha})dt$$

$$\Psi_{s\beta} = \int (U_{s\beta} - \lambda U_{s\alpha} - \lambda\omega_s\Psi_{s\beta})dt$$

$$U_{s\alpha} = V_{s\alpha} - R_s I_{s\alpha}, \quad U_{s\beta} = V_{s\beta} - R_s I_{s\beta}$$

$$\Psi_{smod} = \sqrt{\Psi_{s\alpha}^2 + \Psi_{s\beta}^2}. \tag{2.32}$$

The exciter is modeled to be simplified, as a first-order low-pass filter. The value Ψ_{smod} is computed for each winding, and the mean value is used for control.

The operation at the wind speed of 12 m/s is simulated, with the subsequent increase to 14 m/s and the following decrease to 7 m/s. The change of some variables is displayed in Figure 2.25. The optimal values of the power

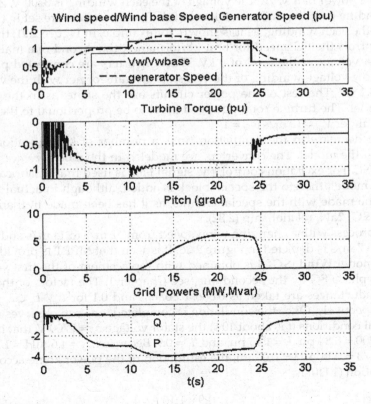

FIGURE 2.25
Processes in the model with six-phase SG.

are reached practically. So, for instance, at the wind speed of 12 m/s, the turbine torque referring to the SG is 0.9 (1 pu of the turbine nominal torque), and the powers of the SG and the network are $0.87 \times 4 = 3.48$ MW and 3.1 MW, respectively, whereas at the wind speed of 7 m/s, they must be by $(7/12)^3$ times lower, that is, they must be equal to 0.69 and 0.615 MW; we have in fact 0.68/0.58 MW, respectively. At the wind speed of 14 m/s, the turbine rotating speed is limited effectively. The scope **Capacitance_Voltage** proves that the voltage across the capacitors in the DC links is maintained to be equal to 6 kV with good accuracy; the scope **Grid** shows that the content of the higher harmonics in the network current is not much: in the nominal condition, THD $\approx$ 3%. Propensity to oscillations can be seen, owing to the limited stiffness of the generator shaft; owing to system configuration, the active damping is not possible, because the diode rectifier transfers the electric energy only in one direction.

The WG sets with nine-phase SG are considered further. The first version of the model of such a generator is employed in the model **Wind_SG_4**. The wind turbine has a power of 7.2 MW at the wind base speed of 12 m/s. The SG has a power of 8 MVA, the voltage of the each winding is 3200 V, $Z_p = 24$, the winding coupling factor is taken as 0.4. Two converters are set in the circuit of the each winding, as it is shown, for example, in Figure 2.21; the first one controls the SG speed for best performance, the second one maintains DC link voltage at the level of 6 kV. Three inverters are connected parallel to the low-voltage winding of the WG step-up transformer with the voltage of 35/3.2 kV. The rest of the power circuits are the same, as in the previous model. The turbine rotating speed is set to be proportional to the wind speed: under $V_w = 12$ m/s $\omega_m^* = 1$.

For a decrease in the simulation time, a number of simplifications are made in the model. The converters are modeled in the mode *average-model-based VSC*. For excitation control, the signal *Flux* is used that is fabricated in the SG model; thereto the special block is added, although in actual fact, it has to be made with the special circuits, as it has been made in the model **Wind_SG_2aN**, relationship (2.32).

The process with a changing wind speed from 12 m/s to 14 m/s and afterward to 7 m/s is depicted in Figure 2.26. It is seen that MPPT is provided.

The model **Wind_SG_5** is obtained with replacement of the first version of nine-phase SG by the second one, see Chapter 1. The factors of the coupling inductances are taken as 0.866, 0.86, 0.1, and 0.1 for L_{m1}, L_{m2}, L_{dq1}, and L_{dq2}, respectively. Once the simulation is completed, one can observes under nominal conditions (t is about 10 s) the stator voltage $v_{st} = 2700$ V, that is, $v_{st} = 2700/3200 = 0.84$ pu, $I_f = 1.45$ pu, and $T_e = 0.9$. Because $L_{md} = 4.6$ mH $= 1.13$ pu, $L_{sle} = L_{sl}/3 = 0.14$ mH $= 0.03$ pu, the estimation of the power angle according to Equation (1.110) is

$$\delta = a\sin\left(\frac{0.9 \times 1.16}{1.45 \times 0.84 \times 1.13}\right) = 49.3°$$

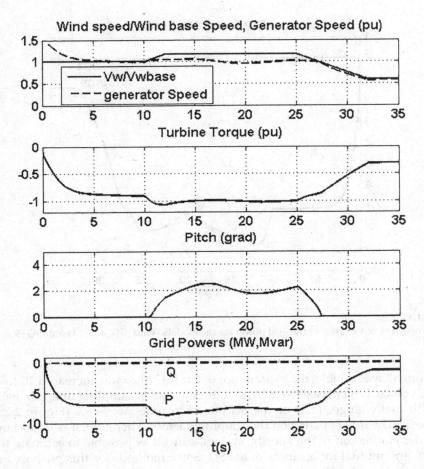

FIGURE 2.26
Processes in the model **Wind_SG_4** with nine-phase SG.

According to **Scope3** *Delta* $\delta = 48°$. The grid powers for both versions are shown in Figure 2.27. It is seen that they are almost the same.

But a large importance for system estimation has content of higher harmonics in its currents. The models **Wind_SG_4a** and **Wind_SG_5a** are the models **Wind_SG_4** and **Wind_SG_5**, in which the mode of the converters *average-model-based VSC* is changed for *switching-function-based VSC*. PWM generators with a carrier frequency of 1620 Hz are added. For the model **Wind_SG_4a**, THD of the SG phase currents are 15–30% (they are different for the windings, least for the middle one) and for the grid current is 2.65%, respectively, whereas these values for the model **Wind_SG_5a** are 7–20% and 2.5%, respectively. Therefore, the parameter alternation of the system components is necessary. The inductance of the SG smoothing reactor

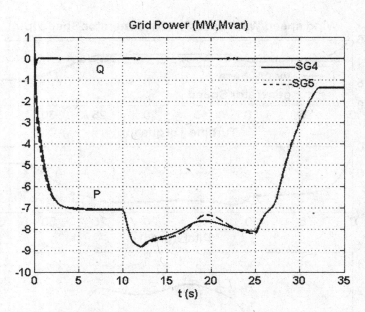

FIGURE 2.27
Comparison of variation of the grid power for the models **Wind_SG_4** and **Wind_SG_5**.

Brunch2 was double, the capacitance of the R-C filter was increased 10-fold, DC voltage is increased to 6.5 kV, and the simulation sampling time is halved. With that change, THD of the SG phase currents are 5–6% (Figure 2.28). The THD value depends on the coupling factors. Therefore, it is important, under simulation of the specific system, as well as possible to estimate the self- and mutual inductances of the SG. Some methods for this purpose are suggested in Ref. [16].

In conclusion of the section, the WG model with the six-phase PMSG, which was described in Chapter 1, is considered, model **Wind_PMSG_4N**. The turbine has the power of 2.5 MW at the base (nominal) wind speed $V_w = 12$ m/s, the PMSG has a power of 3 MVA, its flux linkage $\Psi_r = 6.73$ Wb and $Z_p = 6$. The voltage of the each winding is 4 kV. The rated rotational speed is 400 r/min or 41.87 rad/s, the nominal frequency is $f = 41.87 \times 6/(2 \times \pi) = 40$ Hz. The generator has the interior-mounted magnets, therefore, $L_d \neq L_q$, at this, $L_d = 9$ mH and $L_q = 21.84$ mH. The coupling inductances are taken as $L_{sm} = 3$ mH and $L_{sdq} = 0.78$ mH.

Two back-to-back VSIs are set in the circuits of the each winding; as in Figure 2.21. The VSIs connected to the PMSG control its torque. First, the PMSG maximal power is computed at the actual rotational speed according to Equation (2.7), where P_l is the loss in the stator windings. This power is converted in the reference torque value using the actual rotational speed; this value decreases by the torque loss, which has two terms: constant one and

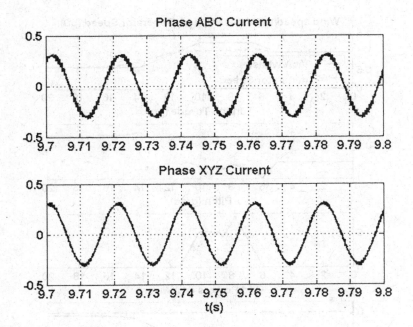

FIGURE 2.28
The SG phase currents.

the one that is proportional to the rotational speed. Utilizing relationships (1.35)–(1.37), the torque reference is recounted in the references for the current components I_q and I_d; the obtained relations are realized with the help of the tables (the program for their calculation is in the field of the option *File/Model Properties/Callbacks*). The current references come to the inputs of the current controllers; the *dq* axes for the controllers of the second winding are shifted by 30° relative to the axes of the first winding.

The VSIs connected to the grid control, as usual, the DC link voltages that are taken as 7 kV, and the reactive power delivered in the grid. These VSIs are connected to the two secondary windings of the transformer having the voltage of 4 kV; the transformer with the primary voltage of 35 kV via the power line and the step-up transformer connects with the network 120 kV.

Variations of the some system quantities are displayed in Figure 2.29 when the wind speed V_w increases from 12 m/s to 15 m/s, with the following drop to 7 m/s. One may observe variation of the rotational speed ω_m proportionally to V_w within the operation range of the latter, and ω_m limitation at $V_w > 12$ m/s. The generator power in this mode is limited also effectual. MPPT is provided: so, at $V_w = 7$ m/s, the generator power must be $(2.5/3) \times (7/12)^3 = 0.165$ pu; actually, according to **Scope**, we have 0.166 pu. It is possible to find with the help of the option *Powerqui/FFT* that THD values for the currents of the generator windings and the grid do not exceed 5%.

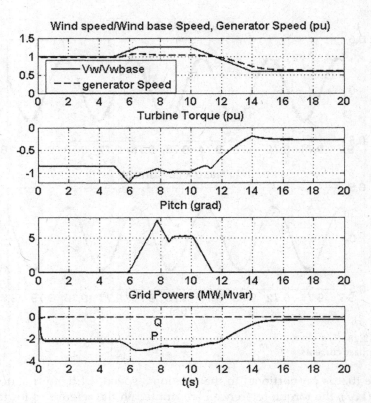

FIGURE 2.29
Variations of the system quantities for the model **Wind_PMSG_4N**.

2.7 Offshore Wind Park with Series-Connected Generators

It has been said already that there is a tendency to use high voltage direct current (HVDC) transmission systems for the offshore parks. To be effective, the transferred voltage has to be rather large, 100–200 kV. The usual configurations with parallel connection require many stages of power conversion, with the necessity to have one or more offshore platforms. The series-connected configuration gives an opportunity to reach a desired voltage level without the very bulky and costly offshore substations.

It can be the various configurations of the offshore parks with series WG connection, Refs. [17–20]. For instance, every WG can have VSI, whose purpose is to control the generator rotational speed by optimal way. With such a structure, when one or some WGs in the series circuit are disabling, the other WGs remain in operation, because there is a path for current in DC link through the bypass diodes of VSI of faulty WGs. But this structure suffers from grave shortcomings. In order that VSI would operate properly, it is necessary that DC voltage was more than the amplitude of the line AC voltage. In this circuit, the DC voltage of the individual VSI is not fixed so that, during the transients, it can drop lower than it is acceptable, which results in shutdown of the given WG, that is, the system turns out unstable. This phenomenon can be seen during simulation of the above-mentioned structure. Therefore, it is not considered further.

One of the main demerits of the structure with series-connected WGs is the high requirement for their insulation relative to the earth. This problem can be solved with utilization of the high- or mid-frequency insulating transformer, Refs. [17, 18].

A type of the DC/AC converter at the receive end is essential in this configuration. When VSI is used, the DC voltage has to be more than the amplitude of the line AC voltage. Therefore, this VSI mission is to keep DC voltage at an appropriate level. When the wind speed for some WGs decreases, their participation in the total DC voltage lessens, and respectively, this participation is more for WGs operating under the higher wind speed. In utmost case, under operational or emergency shutdown of several WGs, the DC voltage of the remainder WG can raise essentially, and such occasion must be anticipated during WG design. On the contrary, DC voltage of the current source inverter (CSI) is less than the line voltage amplitude and, therefore, can change under wind speed variation in accordance with the produced power, if the CSI input current is kept constant. In order not to permit the DC voltage essential decrease under wind speed drop, it is possible to reduce the direct current when the fabricated power lowers noticeably.

The configuration given in Figure 2.30, Ref. [17], is implemented in the model **Wind_PMSG_9N**. The converter **Boost AC/DC** consists of the diode rectifier and DC/DC boost converter (Figure 1.5). The inductor current (diode rectifier load current) is determined as $I_l = P(\omega_r)/U_{dcx}$, where the last quantity is the mean value of the voltage at the rectifier output, $P(\omega_r)$ is the

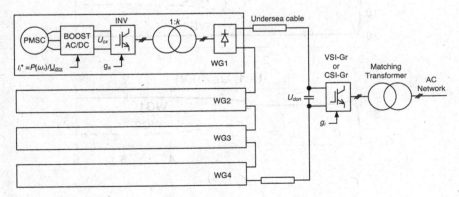

FIGURE 2.30
Offshore configuration with series-connected WGs.

generator power maximal value at the given rotational speed. It was proved under basis of formula (2.31) that MPPT is provided by this way. The additional boost circuit is place at the output of the DC/DC boost converter that consists of the VSI with the modulating frequency of 1000 Hz, the step-up transformer having transformation ratio k and the diode rectifier. By this way, the HVDC input circuits consist of a number of diode bridges that provide a path for the direct current.

Four aggregated blocks are connected in series in the model **Wind_PMSG_9N**; each block has a power of 20 MW under a voltage of 63 kV. Because it is taken $k = 3$, the voltage at the output of DC/DC boost converter is 21 kV. The aggregated generator with a power of 24 MVA has the rotor flux linkage $\Psi_r = 9$ Wb, the pole pair $Z_p = 6$, the nominal rotational speed 60.3 rad/s, and the nominal frequency 57.6 Hz. Because the rms of the line EMF (electromotive force) is $E = 9 \times 6 \times 60.3 \times 1.732/1.4142 = 3988$ V, the voltage at the output of the generator rectifier can be about 5 kV taking the voltage drop across the winding inductances into account.

At the receive end of the power line (on-shore), the three-level VSI keeps the DC line voltage as $63 \times 4 = 252$ kV and delivers the produced electrical energy into the land network via the step-up transformer 120/230 kV.

The process is presented in Figure 2.31 when all WGs have the same wind speed $V_w = 12$ m/s, but V_w of the WG2 drops by 4 m/s at $t = 2$ s. One may see

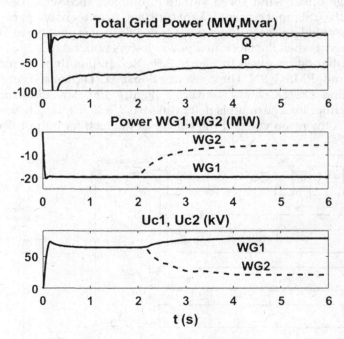

FIGURE 2.31
Processes in the model **Wind_PMSG_9N** under decrease V_w for WG2.

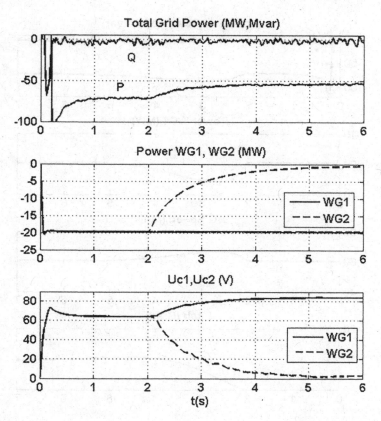

FIGURE 2.32
Processes in the model **Wind_PMSG_9N** under WG2 shutdown.

that the farm remains in operation and WG2 power reduces about to 6 MW, as it must be 20 MW × (8/12)^3 = 5.9 MW. The power delivered in the grid decreases accordingly. DC voltage at the output of the diode rectifier of WG2 reduces, and the DC voltages of the other WGs increase, respectively.

The process is displayed in Figure 2.32 when WG2 stops at t = 2 s. The power delivered in the grid reduces by 25%, but the wind park goes on with its work. The process is depicted in Figure 2.33, when V_w of all WGs arise from 12 m/s to 15 m/s at t = 2 s. It is seen that after transient that lasts 2 s, the system parameters come back to the initial values with angle β increase to 10°.

The three-level inverter is replaced by the modular multilevel converter (MMC) in the models **Wind_PMSG_9M, Wind_PMSG_9Ma,** which was described in Chapter 1. The number of modules in the semi-phase is 10, the voltage of the each module is 25,200 V respectively; it means that the each module model consists of 20–25 actual modules. The inverter is connected to the grid 120 kV via a reactor, without transformer. The control system of the inverter proper (i.e., conversion of the modulating signal into control pulses)

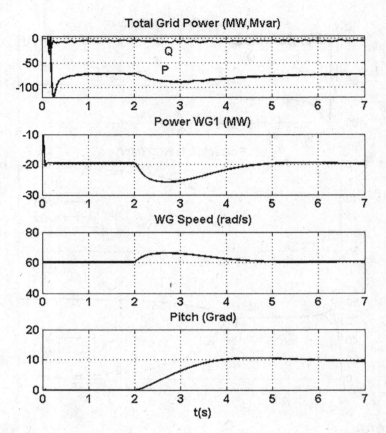

FIGURE 2.33
Processes in the model **Wind_PMSG_9N** under V_w increase.

is described in Chapter 1 (Figure 1.18). The fabrication of the modulating signal is carried out by the following method.

The inverter connects to the grid through the units, which can be accepted as the true inductive ones with the inductance L. The phasor of the grid voltage U_2 has the frequency of 50 Hz; its direction can be taken as a reference. The phasor of the inverter voltage U_1 has the same frequency but is shifted by the angle β. Then the current sent to the grid is

$$I = \frac{U_1 e^{j\beta} - U_2}{jX} = \frac{j(U_2 - U_1 \cos\beta) + U_1 \sin\beta}{X}, \qquad (2.33)$$

where $X = \omega L$. Because the value of β is small usually, the values of the active and reactive powers delivered to the grid may be written as

$$P = \frac{U_1 U_2}{X}\beta \approx \frac{U_2^2}{X}\beta, \qquad Q \approx \frac{U_2(U_2 - U_1)}{X}. \qquad (2.34)$$

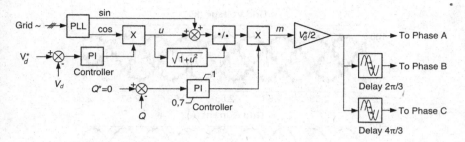

FIGURE 2.34
Block diagram of the control signal fabrication for the model **Wind_PMSG_9M**.

Therefore, the active power can be controlled by change of the inverter phasor angle relative to grid voltage phasor, and the reactive power can be controlled by change of the value the former phasor about the latter.

Implementation of this control method is shown in Figure 2.34. The signals sin ωt and cos ωt are formed with the help of PLL, which are determined by the position of the grid voltage phasor; the first signal is in phase with phasor. The second signal is used to shift the former by the angle β; it is modulated in amplitude by the output of the controller of the DC voltage in the transmission line. In this way, an equality of the active powers, the produced and delivered in the grid, is provided. In turn, the derived signal is modulated in amplitude by the output of the reactive power controller. After multiplication by $V_d/2$ and shifts by 0, $2\pi/3$ and $4\pi/3$, respectively, the signals are fabricated that correspond to the second term in Equation (1.45).

Because simulation in this model runs slowly, only a process segment is displayed in Figure 2.35 when, under an initial wind speed of 12 m/s, it drops to 8 m/s at $ft = 3$ s for all WGs. One can see that the waveforms of the grid voltage and current are satisfactory; the power delivered in the grid corresponds to the optimal values under given wind speeds.

In the model **Wind_PMSG_9Ma**, instead of the separate blocks forming inverter phase, the block **Half-Bridge MMC** is employed that models a semiphase. This block is available beginning from the version R2016b.

The CSI with thyristor is utilized in the model **Wind_PMSG_9M1**; a connection diagram is shown in Figure 2.36. The inverter connects to the grid 735 kV via the three-winding step-up transformer, the low-voltage winding voltage is 120 kV. The control system governs the line current. The maximal value of the current is accepted as $20 \times 10^6 \times 4$ W/252,000 V $\approx$ 320 A. When the wind speed drops, the line voltage decreases as well. In order to avoid a considerable voltage drop, the current reference reduces in accordance with the reducing power, when the generated power drops to 25% of the rated value. The process is depicted in Figure 2.37 when, at the initial wind speed of 12 m/s, it reduces to 8 m/s for all WGs. It is seen that, owing to reducing current, the line voltage decreases only to 100 kV, whereas with the constant current it has to drop to $252 \times (8/12)^3 = 74.7$ kV.

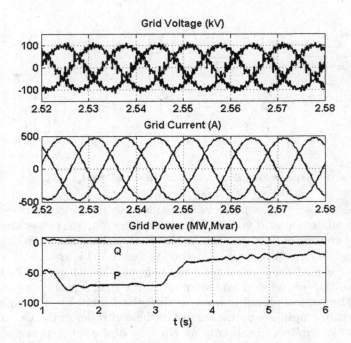

FIGURE 2.35
Processes in the model **Wind_PMSG_9M**.

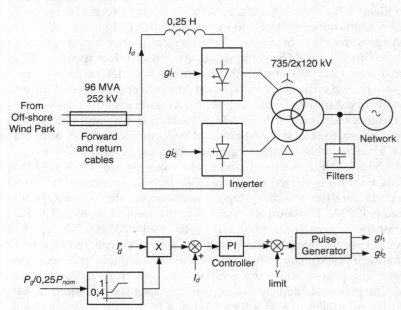

FIGURE 2.36
Connection diagram of the offshore wind park with onshore CSI.

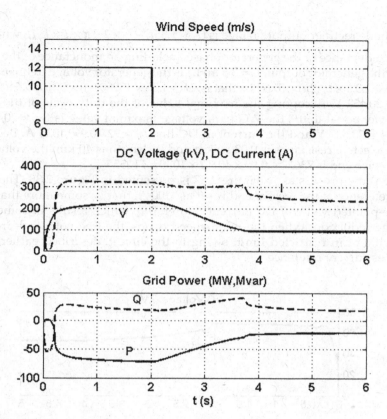

FIGURE 2.37
Process in the model **Wind_PMSG_9M1** under V_w drop.

The systems without galvanic isolation are considered in a number of investigations. The configuration with WG connection in series without the intermediary transformers is suggested in Ref. [19]. Referring to Figure 2.30, it means that the inverter **Inv**, the transformer 1:k, and the diode bridge are omitted. The buck converter is used instead of the AC/DC boost converter, because its output circuits permit the series connection of the analogous units, ensuring that a path always exists for the flow of HVDC link current. MPPT is obtained like in the previous model (Figure 2.30).

Four modules are connected in series in the model **Wind_PMSG_9N1**; each module models four WGs with a power of 5 MW connected in series. The generator parameters are the same as in the previous model; specifically, the nominal rotational frequency is 60.3 rad/s, $Z_p = 6$, so that the line EMF rms value of the equivalent generator is $E = 9 \times 4 \times 6 \times 60.3 \times 1.73/1.41 = 15,980$ V. The rms value of the generator current may be estimated as $I_{grms} = 20 \times 10^6/1.73/15,980 = 723$ A, then the diode rectifier output current is $I_r = \pi \times 723/\sqrt{6} = 927$ A.

The diode rectifier output voltage is $U_r = 1.35E - \dfrac{3}{\pi} \times 2 \times \pi \times f \times L_f I_r$, where L_f is the inductance of the generator phase, including an inductance of the reactor at the generator output, $L_f = 3.5$ mH, f is the generator voltage frequency, in the case $60.3 \times 6/(2\pi) = 57.6$ Hz, that is, $U_r = 1.35 \times 15980 - 6 \times 57.6 \times 3.5 \times 10^{-3} \times 927 = 20{,}452$ V. Accepting the maximal value of the duty cycle of the buck converter equal to 0.9, the WG park voltage maximal value is $U_{dc} = 20{,}452 \times 0.9 \times 4 = 73{,}627$ V and the current in DC line $I_{dc} = 927/0.9 = 1030$ A. Because the line active resistance is 0.03 Ω/km and its length is 40 km, the voltage at the receive end is $73627 - 0.03 \times 40 \times 2 \times 1030 = 71155$ V.

The same process as in Figure 2.37 is presented in Figure 2.38. The current reference at the receive end is set as 1100 A, that is, some more than the above-pointed value of DC current I_{dc}, Ref. [19], which results in some line voltage reduction. When the wind speed drops, the line voltage decreases as well, but in restricted limit, owing to the circuit, described earlier, that reduces current reference.

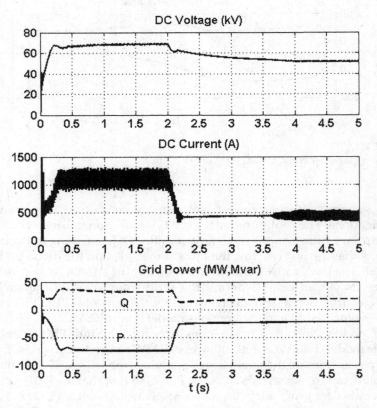

FIGURE 2.38
Process in the model **Wind_PMSG_9N1** under V_w drop.

2.8 Simulation of Two-Terminal Offshore Wind Park

Since several offshore wind farms might locate in a vast sea area and be connected to the different onshore power grids, the multiterminal HVDC system with an offshore DC grid can provide more economic and technical benefits than several individual two-terminal HVDC systems. The offshore park that can transfer an energy produced in two unconnected electric power systems, which can have the different voltages, is considered further. Both receive ends are equipped with the voltage source converters (VSCs). Two different operating modes can be implemented in such a system, Ref. [21]:

1. One network has priority in terms of transmitting power over the other. The second VSC will not receive any power until the capability of the first VSC has been reached or the needs of the first network will be satisfied. For example, the first network has a power shortage that must be eliminated in any possibility; only when this system receives the additional power in sufficient amount, its surplus may be sent to the second network.

2. The both VSCs share a certain amount of power being generated by the wind farm.

Because the DC voltage keeps by the onshore inverters, their control systems contain the voltage controllers with the innermost current controllers, see previous models. In the system, in which the first operating mode is implemented, up limit of the voltage controller of the first VSC (the maximal direct current I_{dc1}^*) is set to be fit with the maximal power of this system under the reference value of DC voltage. When, with wind speed increase, the generated power exceeds this maximal power, the voltage controller will come in saturation, and DC voltage begins to rise. The reference of the voltage controller of the second VSC U_{dc2}^* is accepted more, than the reference voltage of the first VSC U_{dc1}^*, by the voltage drop in the first line under the maximal allowable current:

$$U_{dc2}^* = U_{dc1}^* + R_1 I_{dc1}^* - R_2 I_{dc2}^0, \tag{2.35}$$

where R_1 and R_2 are the resistances of the DC transmission lines including return, I_{dc2}^0 is the minimal current of the second network to keep it in the working state. When the DC rising voltage reaches value U_{dc2}^*, the controller of the second network becomes active.

In the model **Wind_PMSG_8N**, the subsystem WG models the wind park with the power of 250 MW under the wind speed of 12 m/s; the configuration with the low-frequency transformer and VSI-Ge is utilized in WGs (Figure 2.39). The equivalent generator has main parameters as in the previous

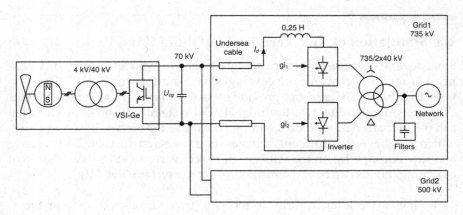

FIGURE 2.39
Block diagram of the two-terminal offshore wind park.

models: $\Psi_r = 9$ Wb, $Z_p = 6$, $\omega_{nom} = 60.3$ rad/s. VSI-Ge carries out the control of the generator rotational speed; its reference is computed by Equation (2.9).

WG park is connected with the onshore thyristor inverter (CSI1) and with the grid 735 kV by the DC line having a length of 40 km and with CSI2 and with the grid 500 kV by 60 km length. The diagrams of the inverters and grids are the same as in the previous model. The DC nominal voltage is 70 kV, the line resistivity is 0.015 Ω/km. The maximal power sent to CSI1 is $P_{1max} = 160$ MW. Therefore, $I_{dc1}^* = 160 \times 10^6 / 70 \times 10^3 = 2286$ A. CSI2 voltage controller reference is

$$U_{dc2}^* \approx 70 + 0.015 \times 40 \times 2 \times 2286 = 72.7 \approx 73 \text{ kV because } I_{dc2}^0 = 50A \approx 0.$$

The process is simulated when, at the initial wind speed of 10 m/s that gives WG power $(10/12)^3 \times 250 = 145$ MW, it rises to 12 m/s at $t = 4$ s. The process is presented in Figure 2.40. It is seen that, with an increase in wind speed, the DC voltage remains limited. The CSI1 power is limited as well, power surplus is sent to CSI2.

To implement the second operating mode, the droop characteristics are used. The inverter DC voltage references are determined as $U_{dc1}^* = U_{dc}^* + K_{r1}I_{dc1}$, $U_{dc2}^* = U_{dc}^* + K_{r2}I_{dc2}$. If V_{dc} is the DC voltage at the sending end, and supposing that the actual voltage is equal to the reference, it follows from relationship

$$V_{dc} = R_1 I_{dc1} + U_{dc}^* + K_{r1}I_{dc1} = R_2 I_{dc2} + U_{dc}^* + K_{r2}I_{dc2} \qquad (2.36)$$

that

$$\frac{I_{dc1}}{I_{dc2}} = \frac{R_2 + K_{r2}}{R_1 + K_{r1}} = n. \qquad (2.37)$$

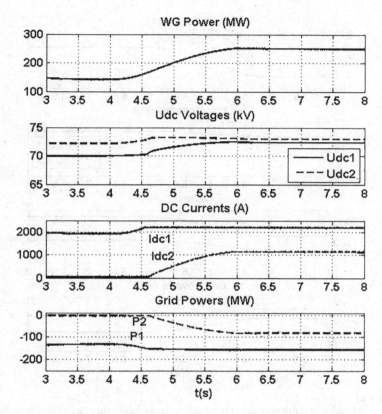

FIGURE 2.40
Process in two-terminal offshore wind park in the first operating mode.

Because the DC voltages of the both systems are nearly equal, the transmitted powers will be in the same ratio.

It is accepted $n = 2$ in the model **Wind_PMSG_8Na**; it means that 2/3 of the produced power goes in CSI1 and 1/3 in CSI2. It follows from Equation (2.37) $0.015 \times 60 \times 2 + K_{r2} = (0.015 \times 40 \times 2 + K_{r1}) \times 2$ or $K_{r2} = 0.6 + 2 K_{r1}$.

The droop factor K_{r1} is taken so that a step-down of the DC voltage is 5%, when the maximal power of 170 MW is transmitted, that is,

$$K_{r1} = \frac{0.05 U_{dc}}{I_{dc}} = \frac{0.05 U_{dc}^2}{I_{dc} U_{dc}} = \frac{0.05 \times 70^2 \times 10^6}{170 \times 10^6} = 1.44,$$

then $K_{r2} = 3.48$. The resulting process is depicted in Figure 2.41. It is seen that, at the wind speeds of 10 m/s and 12 m/s, CSI1 powers are 100 MW and 160 MW, respectively, whereas CSI2 powers are 50 MW and 80 MW, that is, the powers are divided in the ratio 2:1 really.

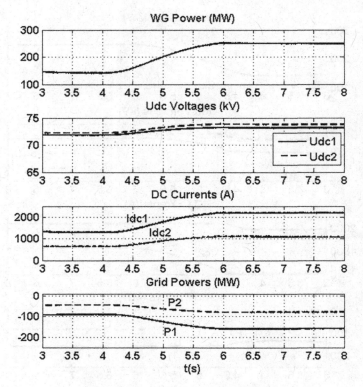

FIGURE 2.41
Process in two-terminal offshore wind park in the second operating mode.

2.9 Direct Power Control of VSI-Gr

In the previously developed models (except for the **Synchronverter** model), as well as in the models considered below, the control of the VSI-Gr is carried out by regulating the components of the current space vector of the network, oriented along the space vector of the network voltage. For this operation, it is necessary to know the angular position of this latter, which is carried out either using PLL or by directly processing the components of this vector, as in the **Wind_IG_6N** model, which can be done with a symmetrical and undistorted network voltage. These calculations, as well as the need for a number of transformations, complicate the control system and slow down the calculations. Therefore, the structures that do not require a coordinate transformation are of interest; this has already been mentioned in Section 2.5.

Besides synchronverter system considered in the referred section, the method direct-power control, developed in Refs. [22, 23] and simulated in this section, is of interest.

The control system in this method operates in a stationary reference frame in the α–β axes. The active and reactive inverter powers are determined as

$$P = 1.5(v_{g\alpha}i_\alpha + v_{g\beta}i_\beta) \tag{2.38}$$

$$Q = 1.5(v_{g\beta}i_\alpha - v_{g\alpha}i_\beta) \tag{2.39}$$

where $v_{g\alpha}$ and $v_{g\beta}$ are the grid voltages, and i_α and i_β are the VSI currents. After differentiation, taking into account the relationships for currents

$$L\frac{di_\alpha}{dt} + Ri_\alpha = u_\alpha - v_{g\alpha} \tag{2.40}$$

$$L\frac{di_\beta}{dt} + Ri_\beta = u_\beta - v_{g\beta} \tag{2.41}$$

where u_α and u_β are the VSI voltages, and having in mind that $dv_{g\alpha}/dt = -\omega v_{g\beta}$, $dv_{g\beta}/dt = \omega v_{g\alpha}$, we obtain

$$\frac{dP}{dt} = -\frac{R}{L}P - \omega Q + \frac{3}{2L}(u_p - V_g^2) \tag{2.42}$$

$$\frac{dQ}{dt} = -\frac{R}{L}Q + \omega P + \frac{3}{2L}u_q \tag{2.43}$$

where $u_p = v_{g\alpha}u_a + v_{g\beta}u_\beta$, $u_q = v_{g\beta}u_a - v_{g\alpha}u_\beta$ are considered as the control variables. The VSI voltage components are taken to be proportional to the grid voltage. Then

$$u_\alpha = \frac{1}{V_g^2}(v_{g\alpha}u_p + v_{g\beta}u_q) \tag{2.44}$$

$$u_\beta = \frac{1}{V_g^2}(v_{g\beta}u_p - v_{g\alpha}u_q), \tag{2.45}$$

where V_g is the grid phase voltage amplitude. At this, the modulating signals at PWM input are $m_\alpha = \dfrac{2u_\alpha}{U_{dc}}$, $m_\beta = \dfrac{2u_\beta}{U_{dc}}$, where U_{dc} is the VSI DC link voltage.

The variables u_p and u_q contain the feedforward and feedback inputs as

$$u_p = V_g^2 + \frac{2L}{3}(\omega Q + u_{pr}) \tag{2.46}$$

$$u_q = \frac{2L}{3}(u_{qr} - \omega P) \tag{2.47}$$

where u_{pr} and u_{qr} are the outputs of proportional integral (PI) controllers of the active and reactive powers, respectively. For VSI-Gr, the active power reference is determined by the output of the DC link voltage controller.

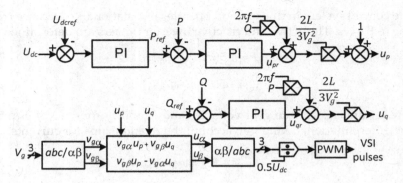

FIGURE 2.42
Block diagram of direct power control.

Thereby we come to the block diagram of the control system shown in Figure 2.42. Such a system is simulated in the model **Wind_PMSG_1Na** that is the model **Wind_PMSG_1N**, in which the model **Drive train** is removed and the VSI-Gr control system is replaced. The wind speed, initially equal to 10 m/s, increases to 12 m/s with the rate of 1 m/s² at $t = 5$ s, and then again decreases to 10 m/s at $t = 10$ s. The process is shown in Figure 2.43. It can be seen that the MPPT is implemented, the reactive power is close to zero, and the DC voltage is kept constant.

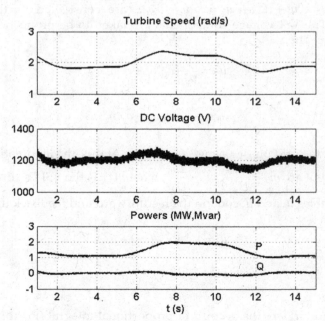

FIGURE 2.43
Processes in the model **Wind_PMSG_1Na**.

References

1. Global Wind Energy Council (GWEC). Global Wind Report 2017. GWEC, Brussels, 2018, April.
2. Ragheb, M. Modern wind generators. NPRE, 475(8), 2018, Marth.
3. Ackermann, T. (ed.). Wind Power in Power Systems. John Wiley & Sons, Ltd, West Sussex, 2005.
4. Perelmuter, V. Renewable Energy Systems, Simulation with Simulink® and SimPowerSystems. CRC Press, Boca Raton, FL, 2017.
5. Dolan, D. S. L., and Lehn, P. W. Simulation model of wind turbine 3p torque oscillations due to wind shear and tower shadow. IEEE Transactions on Energy Conversion, 21(3), 2006, September, 717–724.
6. Muyeen, S. M., Hasan Ali, Md., Takahashi, R., Murata, T., Tamura, J., Tomaki, Y., Sakahara, A., and Sasano, E. Comparative study on transient stability analysis of wind turbine generator system using different drive train models. IET Renewable Power Generation, 1(2), 2007, June, 131–141.
7. De Kooning, J. D. M., Vandoorn, T. L., Van de Vyver, J., Meersman, B., and Vandevelde, L. Shaft speed ripples in wind turbines caused by tower shadow and wind shear. IET Renewable Power Generation, 8(2), 2014, 195–202.
8. Geng, H., Xu, D., Wu, B., and Yang, G. Active damping for PMSG-based WECS with DC-link current estimation. IEEE Transactions on Industrial Electronics, 58(4), 2011, April, 1110–1119.
9. Li, Y., Xu, Z., and Wong, K. P. Advanced control strategies of PMSG-based wind turbines for system inertia support. IEEE Transactions on Power Systems, 32(4), 2017, July, 3027–3037.
10. Burra, R., Ambekar, A., Narang, H., Liu, E., Mehendale, C., Thirer, L., Longtin, K., Shah, M., and Miller, N. GE brilliant wind farms. IEEE Symposium on Power Electronics and Machines for Wind and Water Applications, 2014, 1–10.
11. Miller, N. W. GE Experience with Turbine Integrated Battery Energy Storage. www.ieee-pes.org/presentations/gm2014/PESGM2014P-000717.pdf
12. Zhong, Q.-C., Nguyen, P.-L., Ma, Z., and Sheng, W. Self-synchronized synchronverters: Inverters without a dedicated synchronization unit. IEEE Transactions on Power Electronics, 29(2), 2014, February, 617–630.
13. Zhong, Q.-C., and Weiss, G. Synchronverters: Inverters that mimic synchronous generators. IEEE Transactions on Industrial Electronics, 58(4), 2011, April, 1259–1267.
14. Bose, B. K. Modern Power Electronics and AC Drives. Prentice Hall PTR, Upper Saddle River, NJ, 2002.
15. Hinkkanen, M., and Luomi, J. Modified integrator for voltage model flux estimation of induction motors. IEEE Transactions on Industrial Electronics, 50(4), 2003, August, 818–820.
16. Tessarolo, A., Mohamadian, S., and Bortolozzi. M. A new method for determining the leakage inductances of a nine-phase synchronous machine from no-load and short-circuit tests. IEEE Transactions on Energy Conversion, 30(4), 2015, December, 1515–1527.
17. Molinas, M. Offshore wind farm research: Quo vadis? Wind Seminar 2011, Norwegian University of Science and Technology (NTNU), Trondheim, 2011, 1–47.

18. Holtsmark, N., Bahirat, H. J., Molinas, M., Mork, B. A., and Høidalen, H. Kr. An all-DC offshore wind farm with series-connected turbines: An alternative to the classical parallel AC model? IEEE Transactions on Industrial Electronics, 60(6), 2013, June, 2420–2428.
19. Veilleux, E., and Lehn, P. W. Interconnection of direct-drive wind turbines using a series-connected DC grid. IEEE Transactions on Sustainable Energy, 5(1), 2014, January, 139–147.
20. Qiang Wei, Q., Wu, B., Xu, D., and Navid Reza Zargari, N. R. Power balancing investigation of grid-side series-connected current source inverters in wind conversion systems. IEEE Transactions on Industrial Electronics, 64(12), 2017, December 9451–9460.
21. Xu, L., Williams, B. W., and Yao, L. Multi-terminal DC transmission systems for connecting large offshore wind farms. IEEE Power and Energy Society General Meeting—Conversion and Delivery of Electrical Energy in the 21st Century, IEEE, Pittsburgh, PA, 2008, 1–7.
22. Gui, Y., Kim, C., and Chung, C. C. Grid voltage modulated direct power control for grid connected voltage source inverters. Proceeding of the American Control Conference, 2017, 2078–2084.
23. Gui, Y., Wang, X., Blaabjerg, F., and Pan, D. Control of grid-connected voltage-source converters. IEEE Industrial Electronics Magazine, 13(2), 2019, June, 31–40.

3

Photovoltaic Energy Sources

3.1 Fundamentals

Photovoltaic (PV) energy sources fabricate electricity by direct transformation of the solar light. The photocurrent value produced by one PV solar cell is low, so that these cells are arranged in modules and arrays of modules. Equivalent circuit of a PV cell (PC) is depicted in Figure 3.1.

The diode current is determined by relationship

$$I_d = I_{d0} \left\{ \exp\left[\frac{qV_1}{K_b T_c A (1 - b_t (T_c - T_{cr}))} \right] - 1 \right\}, \tag{3.1}$$

where I_{d0} is the diode saturation current, A, $q = 1.602 \times 10^{-19}$ C is the electron charge, $K_b = 1.38 \times 10^{-23}$ J/K is Boltzmann's constant, T_c is the PC absolute temperature, T_{cr} is the nominal temperature, A is the adjusting factor, and b_t is the voltage temperature coefficient.

The magnitude of the photocurrent depends on the irradiance level and on the temperature; this dependence may be taken as

$$I_{ph} = I_{sr}[1 + a_k (T_c - T_{cr})] \frac{Q}{1000}, \tag{3.2}$$

where I_{sr} is the photocurrent under the nominal conditions, when $T_c = T_{cr}$ (298 K usually), and Q is the irradiance level in W/m^2, a_k is the current temperature coefficient.

Resistors R_p and R_s take into account the losses in a PV cell.

As it can be seen, there is a direct connection between the diode voltage (the cell output practically) and the diode current, which, in turn, determines this output voltage. Under simulation, the so-called algebraic loop is formed; its existence makes simulation slower, and one can observe the cases when simulation stops. In the last case, it is necessary to redesign the model. If to try to break the algebraic loop, for instance, by using a one-simulation-step time delay, a distortion of I-V cell characteristics can have place under the voltages, in the vicinity of the maximum point and larger; it will be shown further.

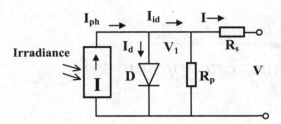

FIGURE 3.1
Equivalent circuit of a PV cell.

The photocell model (PV) was developed by the author, Ref. [1], which implements the relationships (3.1) and (3.2), and the schematic is shown in Figure 3.1. In the model's dialog box, the parameters included in these relationships, and shown in Figure 3.1, have to be determined, as well as the number of series and parallel connected PV cells. This model was employed successfully in a number of models in Ref. [1].

Beginning from R2015a version, the PV array model is included in SimPowerSystems that uses Equation (3.1) with a slight change. Over 10,000 PV modules from many manufacturer datasheets can be preset. Besides, PV module with user-defined parameters can be specified. In the first case, the number of PV modules connected in series in each string and the number of strings of series-connected modules that are connected in parallel must be assigned solely. In the second case, the parameters such as the open circuit voltage V_{oc}, the voltage at maximum power point V_{mp}, the short-circuit current I_{sc} and the current at maximum power point I_{mp}, the number of cells per module N_{cell}, as well as the temperature coefficients for the open circuit voltage and the short-circuit current have to be entered. When these quantities are defined, the block computes the parameters entering in Equations (3.1) and (3.2). Besides, the I-V and P-V characteristics of one module or of the whole array can be displayed.

In the second page of the dialog box, there is an opportunity to select the option *Break algebraic loop in internal model*. It has already been mentioned that this option must be chosen with caution.

Of course, the employment of this model makes the preparation for the simulation much easier, but it can be observed during simulation that it runs slower than with use of the author's model, so that when one simulates a complicated system, one can try both models to determine the most suitable.

The models **Photo_m1** and **Photo_m1N** are intended for investigation of characteristics of the PV model. The PV cells supply the variable resistor, whose resistance smoothly changes from $0.0039 \times R$ *ns/np* Ω to R *ns/np* Ω by 255 levels and afterward to infinity. Here *ns* and *np* are the numbers of the cells connected in series and in parallel, respectively. The resistance block consists of eight resistors, whose resistances change as a power of 2, and which are switched in parallel. The coding unit converts the signal 0–255 in the binary code that controls the switches.

Under the experiments, $ns = np = 1$ were set in the model **Photo_m1** initially, and the parameters of one PV cell were found as $V_{oc} = 0.644$ V, $V_{mp} = 0.53$ V, $I_{mp} = 7.4$ A, and $I_{sc} = 8.03$ A. Afterward, these parameters are entered in the dialog box of the model **Photo_m1N** (the voltages multiplied by $N_{cell} = 60$); moreover, it is taken the number of series-connected modules per string equal to 14 and the number the parallel strings equal to 50. It is accepted in the model **Photo_m1** the number of parallel PCs = 50 and the number of series PCs = 14 × 60 = 840. Dependences of I and P on V for both models are depicted in Figure 3.2. It is seen that they are the same. Therefore, both models can be used under simulation of the electrical systems. It can be observed when performing simulation that the model **Photo_m1** runs noticeably faster than another one. It should be noted that, for the first model, the irradiance is given in pu and the temperature in Kelvin scale, whereas for the second model in (W/m²) and Celsius, respectively.

The model **Photo_m1N** was simulated with the algebraic loop. If to repeat simulation with the broken loop, the considerable characteristic distortion can be seen.

As can be seen from Figure 3.2, the dependence of P on V has a maximum, and for gaining the maximum efficiency from an array of PV cells (PV), its voltage must be V_{mp}. In order to realize this requirement, the following

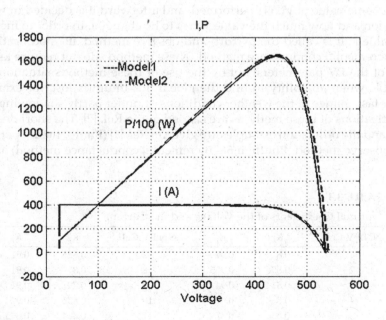

FIGURE 3.2
Comparison of the PV author's model and SimPowerSystems model.

conditions have to be fulfilled: (a) presence of the device that can change V; (b) availability of the unit that can determine the optimal point or deviation from it; (c) availability of the algorithm for V adjusting. As a device that can change V, the DC/DC or DC/AC converters are employed depending on the system structure. The methods for V_{mp} determination and adjusting can be designated as the direct and indirect. The latter methods are based on the fact that the ratios $K_v = V_{mp}/V_{oc}$ and $K_i = I_{mp}/I_{sc}$ for the certain PV cell type are almost constant when environments change. If the values K_v or K_i are known by means of computations, simulation, or experiment, and the PV cell open circuit voltage or the PV cell output short-circuit current can be measured, the reference quantities of the voltage or the current may be obtained, which can be realized with the aid of the feedback control system.

In Table 3.1 the values of the coefficients K_v and K_i are given under the nominal condition for some selected PV cells, which are included in the PV array model.

The values V_{oc} and I_{sc} can be obtained in two ways: either to place the special measuring PV cells of the same type as the working ones inside the PV array or to switch over the output array circuits. The additional PV cells are necessary in the first case and the switching devices in the second case; moreover, the load loses supply in the measuring moments. The first variant, with V_{oc} or I_{sc} measurements, is realized in the models **Photo_m4V** and **Photo_m4I**, respectively, Ref. [1].

In the direct methods, the measurement of the PV array output power $P(k)$ under some value of $V(k)$ is performed, and afterward it is decided in which direction and how much the value V has to be changed, in order to increase the value P. It is called the perturb-and-observe method. Its modification is the incremental conductance method. These methods do not require knowledge of the PV parameters, but by the use of these methods situations are possible when an ability to find the global maximum is lost, for example, under fast change of the weather conditions or under partly shadowing. The investigations of these methods are carried out in Ref. [2]. The short descriptions and the utilization examples are given in Ref. [1] (**Photo_m4S**—perturb-and-observe method, **Photo_m5**—incremental conductance method) and in

TABLE 3.1

Optimal Coefficients of the Voltage and the Current

NN PV Cell	K_v	K_i	NN PV Cell	K_v	K_i
1	0.8	0.94	7	0.825	0.91
2	0.806	0.935	8	0.8	0.91
3	0.81	0.94	9	0.79	0.91
4	0.8	0.916	10	0.835	0.92
5	0.83	0.87	11	0.824	0.895
6	0.83	0.925	12	0.827	0.937

Ref. [3] (**250-kW Grid-Connected PV Array**—perturb-and-observe method, **Detailed Model of a 100-kW Grid-Connected PV Array**—incremental conductance method).

The operations of PV in the systems for the electrical energy production are simulated further. At that, there is no significance often, what maximum power point tracking (MPPT) method is utilized; suffice it to know that one of them is implemented. So, in the models given here, MPPT is employed based on V_{oc} measurement by the special measuring PCs and on the knowledge of coefficient K_v. Four measuring PCs are supposed loaded with the resistors having the large resistance, so that $V_{mp}^* = K_v V_{oc} n_s / 4$, where n_s is the number of the PVs in the array connected in series.

In conclusion, it is worth noting that the circumstantial overview of the present MPPT methods is given in Ref. [4].

3.2 Simulation of Grid-Connected Photovoltaic Systems

Grid-connected three-phase PV systems have a power from some hundreds to some thousand kilowatts. They can have the different configurations. So, all the strings of PV array can be connected in parallel with their DC outputs, and one VSI is used for DC/AC conversion (Figure 3.3a). It is possible only when DC voltage surpasses the amplitude of the grid phase-to-phase voltage V_{gamp}. If it is the case, the inverter output current (its I_d component) is controlled to keep V_{mp} voltage at the PV output. Such a system is simulated in the model **250-kW Grid-Connected PV Array**, Ref. [3], where $V_{mp} \approx 500$ V and $V_{gamp} \approx 355$ V. Analogously, it is taken $V_{mp} \approx 430$ V and $V_{gamp} \approx 310$ V in the model **Photo_m8** in Ref. [1].

If DC voltage is not more than V_{gamp}, two-stage conversion is employed. The first stage is the DC/DC boost converter, and the second one is VSI. The former keeps the PV array output voltage as V_{mp}, whereas the latter maintains the given reference voltage U_{dc} at the DC/DC converter output. Such a system is simulated in the model **Detailed Model of a 100-kW Grid-Connected PV Array**, Ref. [3], where $V_{mp} \approx 273$ V, $U_{dc} = 500$ V, and $V_{gamp} \approx 367$ V.

The considered structure has the disadvantage that the individual PCs can have the different irradiance, and therefore each PV module may not operate at its maximum power point, which results in less energy harvested.

In order to overcome this demerit, the total PV array is broken up into PV groups, down to the separate strings, which, with the high degree of probability, are under the same environmental conditions; every group is provided with the individual, one- or two-stage converter. Because the voltage of each group is not high usually, the utilization of the DC/DC boost converter seems indispensable. Then it is possible either to connect their outputs in parallel to one VSI (Figure 3.3b) or to have for each group individual VSI (Figure 3.3c).

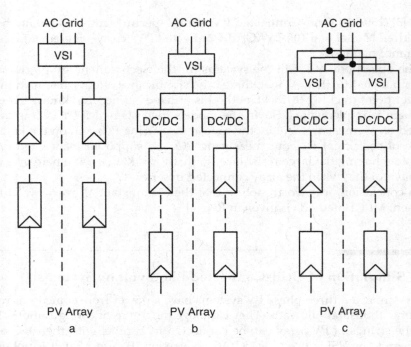

FIGURE 3.3
Possible configurations of the grid-connected three-phase PV systems: (a) one-step conversion, (b) two-step conversion, and (c) two-step conversion with individual VSIs.

The first option is realized in the model **400-kW Grid-Connected PV Farm (Average Model)**, Ref. [3], in which four PV arrays, each has the power of 100 kW, are connected in parallel via DC/DC boost converters to one VSI. Because it is impossible to observe the actual waveforms of the voltages and currents in the average model, the detailed model **Photo_9N** is developed. In this, the PV models discussed in Ref. [1] are used. Three PV arrays, each has the power of 300 kW, are connected in parallel with DC/DC boost converters to a three-level VSI. PV arrays function under different environmental conditions, and the considered modes are shown in Figure 3.4. VSI, by means of the L-C filter and step-up transformer, is connected to the load 200 kW and the grid 35 kV. The VSI control system keeps DC link voltage equal to 500 V and zero reactive power. The variations of the power delivered in the grid under different environmental conditions are displayed in Figure 3.5.

The PV model discussed in Ref. [1] is replaced with the PV model from the version 2015a SimPowerSystems in the model **Photo_9N1**, whose parameters are chosen close to the former one. The model consists of $180 \times 7 = 1260$ strings, each have the power of 255.7 W. One such a string is employed as a measuring one. The process obtained during simulation is shown in Figure 3.6. In versions R2016b and R2019a, in the acceleration mode, the models **Photo_9N_N** and **Photo_9N1_N** have to be used, respectively, instead.

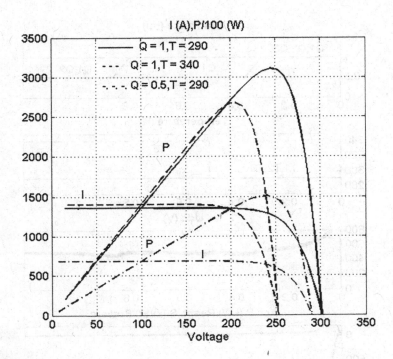

FIGURE 3.4
PV characteristics under different environmental conditions in the model **Photo_9N**.

The configuration considered has a demerit that with outage of the only one system unit, the inverter, the whole set stops energy production. Besides, if an increase in the set power with addition of the extra strings is intended, the output inverter has to be replaced. The structure with the individual inverter for every DC/DC converter does not have this disadvantage (Figure 3.3c). Such a system is modeled in the model **Photo_9N2**. When comparing with the model **Photo_9N**, it can be seen that two-level inverters are connected with the outputs of the PV units (PV array and DC/DC converter); the inverter control systems are the same as in the mentioned model. The inverter outputs are connected in parallel to the step-up transformer. Simulation results are displayed in Figure 3.7. One may see that, unlike the previous models, the reactive power sent in the grid is not zero, because in the considered configuration, not the power at the input of the common transformer but the powers at the outputs of the individual inverter reactors, are controlled.

The model **Photo_9N3** is the same as **Photo_9N2**, but the PV model from SimPowerSystems, version R2015a, is used. The processes in the model do not differ practically from the one given in Figure 3.7.

The normal operation modes are considered in the previous models. But investigation of the emergency conditions is of the interest as well. As a

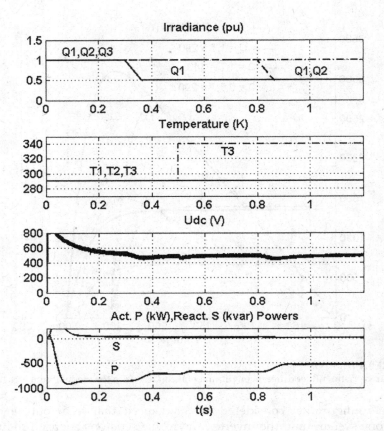

FIGURE 3.5
Processes in the model **Photo_9N**.

fault, the partial short circuit in the grid is considered that PV array senses as a grid voltage sag at point of common coupling (PCC). PV sets have to remain in operation after the fault is cleared. PV action in this situation depends on the array configuration. More intricate effects have a place with a two-stage converter. In this case, the grid inverter tries to send all power produced in the grid under lower grid voltage, which results in an increase of inverter current. Because usually the permissible inverter current does not greatly exceed its rated value, the rise of the inverter current is limited and all energy produced cannot be delivered in the grid that results in uncontrolled increase of DC voltage. The special measures have to be assumed to prevent this effect, Ref. [5]. Specifically, it is necessary to step down the power produced by PV set. It can be obtained with an increase of PV voltage above the optimal value V_{mp}. Such a system is simulated in model **Photo_9N4** which is similar to the above-described model **Photo_9N** in which the 35-kV source with internal impedance is replaced with the

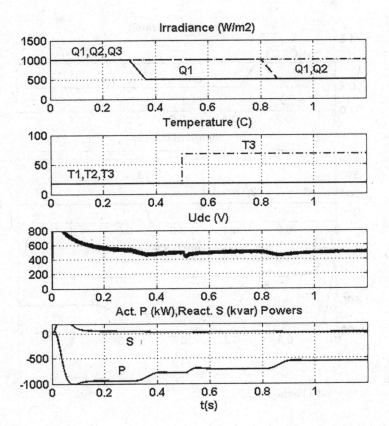

FIGURE 3.6
Processes in the model **Photo_9N1**.

three-phase programmable voltage source with external impedance. The source voltage drops to 50% within 0.6–0.8 s. The circuits are added in the control systems of the individual PV arrays that increase the PV reference voltages when the DC link voltages rise to 520 V with the rated voltage of 500 V; its schematic is depicted in Figure 3.8. When the DC voltage reaches 520 V, the integrator **I** output voltage increases, which results in the rise of the voltage reference signal at the PV output from the value V_{mp}^* to the value, whose maximum is 1.3 V_{mp}^*, which is fit to open circuit. The circuits are activated when steady conditions arise after model starts, in the case, at $t > 0.5$ s.

Grid and DC voltages, VSI current, and grid powers are shown in Figure 3.9. At this, the maximum of I_d was set to 4 kA (under nominal conditions about 3 kA). It is found that DC voltage is limited effectively. It is worth noting that without PV power limitation, DC voltage during grid voltage sag would reach 1.8 kV.

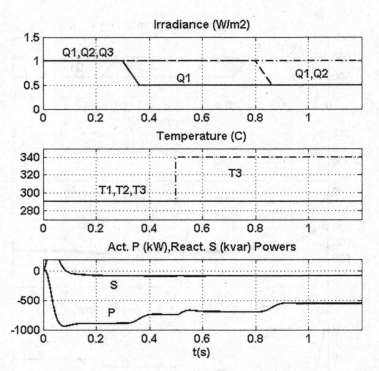

FIGURE 3.7
Processes in the model **Photo_9N2**.

As for the systems with the single-stage converter—VSI, the processes in them, during grid voltage sag, run easier owing to a kind of self-regulation. Under grid voltage sag and inadequate increase of inverter current to transfer all fabricate energy in the grid, the voltage rises at the inverter input; since it is PV output, its power drops in comparison with the previously effective optimal quantity that limits the further voltage growth.

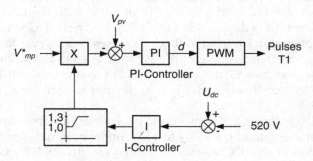

FIGURE 3.8
Circuits for limitation of DC link voltage rise.

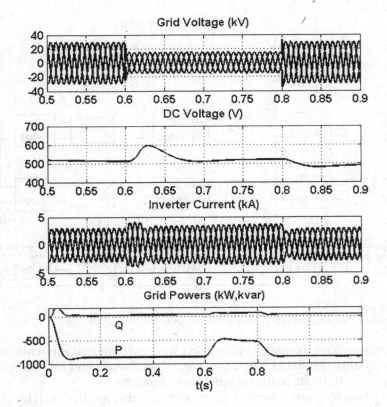

FIGURE 3.9
Processes in the system with two-stage conversion during grid voltage sag.

3.3 Grid-Connected Photovoltaic Systems with Cascaded H-Bridge Multilevel Inverters

Power rise of the PV energy sources promotes employment of the multilevel VSI; thereupon, utilization of the cascaded H-bridge multilevel inverters is very promising (Figure 1.16). Every single-phase bridge of this inverter has to be supplied from an isolated source that is provided in a natural way in the circuits with PV, with supply of every bridge from the separate PV array. The main problem with the use of these inverters is the possible unbalance when the PVs connected to the different bridges produce the different actual powers. One differentiates per-phase (or internal phase) and interphase unbalance. Because all bridges in the phase conduct the same current, per-phase unbalance manifests itself as the different voltages of these bridges, which can reach intolerable values. Phase-to-phase unbalance that appears owing to the difference in the total powers of PVs connected to the bridges of the

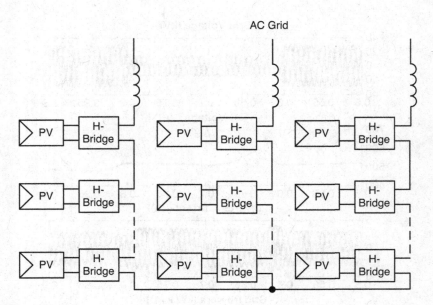

FIGURE 3.10
Configuration with direct connection of the strings to the converter bridges.

phases results in unbalance in the grid phase currents that is not allowed by power quality standards. Therefore, the PV set circuits and the control units purpose to eliminate or to diminish these demerits.

In the simplest case, every string connects to its bridge directly (Figure 3.10), Ref. [6]. It is rather difficult to harvest the maximal power from the separate PV strings and to provide the balanced current injected in the grid in this structure.

In another configuration, all PV strings connect to the bridges not directly but via DC/DC converters. These converters give the possibility to extract the maximal power from PV strings and, as appropriate, to carry out the galvanic isolation (Figure 3.11), Ref. [7, 8].

Such a system is simulated in the model **Photo_Hinv1** (**Photo_Hinv1N_2016** for R2016b, R2019a). The inverter has three H-bridges in the phase (n = 3). Each model of nine PV units with DC/DC converters is the same as in the previous models (for instance, in **Photo_9N4**), that is, it has the power of 300 kW. To speed up simulation that is slow, the circuits for PV output voltage optimal value determination are eliminated; it is supposed that its value, equal to 230 V in the case, is determined by one or another method. The rated value of the VSI DC link U_{dc} = 1000 V. The control system of the VSI output current is the same as in a number of the previous models: the reference for I_d current component is determined by the controller output, whose feedback signal is the sum $\sum_{1}^{3n} U_{dc}$, the reference for I_q current component is determined by the reactive power

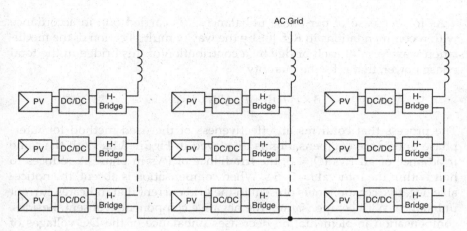

FIGURE 3.11
Configuration with connection of the strings to the bridges via DC/DC converters.

controller. A unipolar modulation is used. The DC voltage reference sum decreases if some voltage U_{dc} exceeds 1200 V. The inverter power via the step-up transformer 2500/6300 V is delivered in the grid and the local load of 1 MW.

The circuits to decrease interphase and per-phase unbalances are provided in the model. A compensation of the former is reached with unbalancing of the phase voltages inversely proportional to the power unbalance of the converter; zero sequence voltage v_0 is added to each reference to introduce the corresponding neutral shift. This sequence does not manifest itself in the inverter line voltage. Specifically, if to denote v_a^*, v_b^*, v_c^*, the outputs of the phase current controllers that are the modulation waveforms in the absence of compensation, with the compensation, the modulation waveforms are taken as

$$v_a = v_a^* - v_0, \; v_b = v_b^* - v_0, \; v_c = v_c^* - v_0, \tag{3.3}$$

where in turn

$$v_0 = 0.5\left[\max(V) + \min(V)\right] \tag{3.4}$$

$$V = v_a^* \frac{P_m}{P_a} + v_b^* \frac{P_m}{P_b} + v_c^* \frac{P_m}{P_c} \tag{3.5}$$

where P_a, P_b, and P_c are the powers of the proper phases, and $P_m = 1/3(P_a + P_b + P_c)$ is the mean power, Ref. [7].

In the considered model, the relationships (3.3)–(3.5) are realized in the subsystem **Balance**, which is active with the tumblers in the low position.

As for decrease of per-phase unbalance, it is carried out, in accordance with recommendations in Ref. [8], by the way of multiplication of the modulation waveform of each bridge by a contribution of this bridge in the total phase power, that is by the quantity

$$3 \times P_{ij}/(P_{i1} + P_{i2} + P_{i3}), i = a, b, c.$$

The process that confirms an effectiveness of the used method for interphase unbalance compensation is displayed in Figure 3.12. With the same irradiance of all PVs ($Q = 1$), the irradiance of PVs in phase A reduces to half within the interval $t = 3–5$ s. When compensation is absent, the noticeable negative component is present in the grid current that indicates current unbalance. With compensation, the negative component is absent. Besides, compensation implementation decreases unbalance of the DC voltages of the phases.

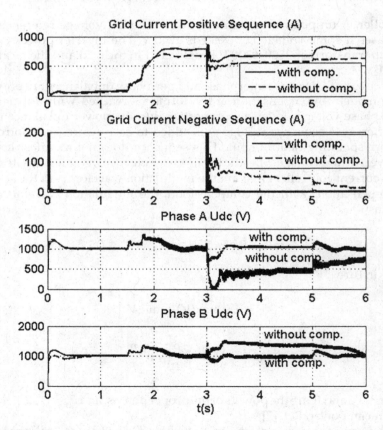

FIGURE 3.12
Effectiveness of interphase unbalance compensation, model **Photo_Hinv1**.

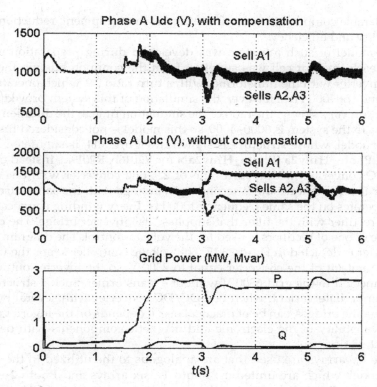

FIGURE 3.13
Effectiveness of per-phase unbalance compensation, model **Photo_Hinv1**.

The voltages at the inputs of the phase A bridges are displayed in Figure 3.13 when the irradiance of the PV2 and PV3 strings halves ($Q_2 = Q_3 = 0.5$) within interval $t = 3–5$ s with the nominal irradiance of PV1 string ($Q_1 = 1$). When compensation is absent, the noticeable unbalance of the DC voltages can be seen; with compensation, the unbalance is absent practically.

The structures considered have the demerit that the balancing circuits affect a modulation coefficient that can limit the power of the PV set or result in distortion of its currents while approaching the nominal conditions. Therefore, the configurations with the DC common bus are proposed, Ref. [9]. DC/DC converters of PVs are connected in parallel and form DC bus with the voltage U_{dc}. These converters provide maximal power harvesting. The voltage U_{dc} is keeping equal to the reference by inverter current control. The DC voltage for each single-phase bridge is fabricated with help of the additional DC/DC converters having galvanic separation input and output. The control systems of these converters keep the given voltage at their outputs, that is, at the bridge inputs. Such a structure provides an extraction of the maximal PV power and inverter balanced current, but at the expense of considerable complication and rise in cost of the equipment, reduction of its reliability and efficiency.

The model of such a system was developed during preparation of this book; each invertor cell was supplied from the common bus via the DC/DC converter with the transformer with a turn ratio 2.9, which operates at a frequency of 1000 Hz. However, the simulation of this system proved that it would run very slowly: the ratio of the simulation time to the duration of the process in the system is 3000–4000, so this model is not considered further.

The model, whose structure is shown in Figure 3.14, is simulated in the model **Photo_Hinv2a** (**Photo_Hinv2aN** for R2016b, R2019a). Instead of nine DC/DC converters with a turn ratio of 2.9, one converter with one transformer is used that has nine secondary windings with the same turn ratio and operates on the same frequency 1000 Hz. Every winding connects to its diode rectifier with the filter that supplies "its" inverter bridge. The output voltage of one of rectifiers is used for the voltage control. The diagram of the converter is depicted in Figure 3.15. The voltage controller keeps the voltage at each output of the converter equal to 2300 V, so the inverter output can be connected to the grid 6300 V without a transformer. Such a structure or its intermediate variant, when one DC/DC converter with several outputs serves some bridges, can be of practical use: it depends on the inverter power and availability of the electronic and magnetic components with demand characteristics.

Nine PV arrays are used that are analogous to the utilized in the previous model, which are united as: **Photo 1**—six arrays and **Photo 2**—three arrays. The DC voltage of parallel-connected converters of PV arrays is kept at 900 V, by control of the active component of the inverter output current, like in the previous model. The process is shown in Figure 3.16, when the irradiance of six arrays drops from 1 to 0.5 at $t = 5$ s. It results in power

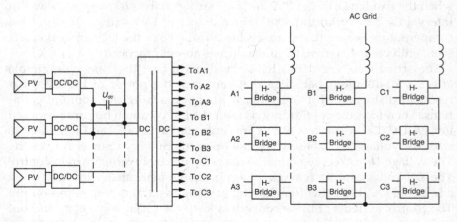

FIGURE 3.14
Configuration of the system with the common transformer for galvanic separation.

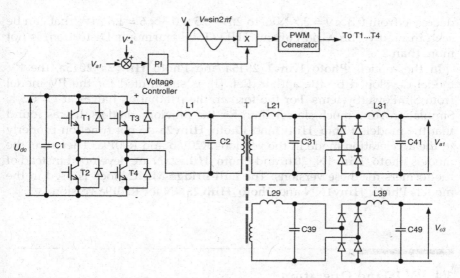

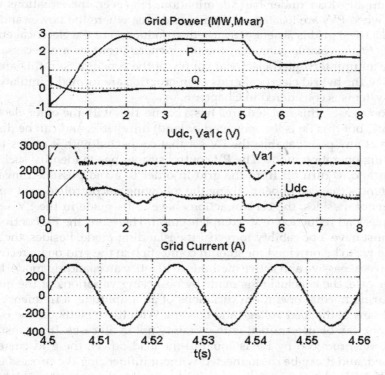

FIGURE 3.15
Circuits of the DC/DC converter with multiple outputs.

FIGURE 3.16
Processes in the model **Photo_Hinv2a**.

decrease from $0.3 \times 9 = 2.7$ MW to $2.7 - 0.5 \times 0.3 \times 6 = 1.8$ MW that can be seen in Figure 3.16. Grid current THD (Total Harmonic Distortion) is not more than 2%.

In the models **Photo_Hinv1_2015a** and **Photo_Hinv2a_2015a**, the PV model developed by the author, Ref. [1], is substituted for the PV model from SimPowerSystems. For the former, the irradiation halves at $t = 0.3$ s. Simulation runs much slower than for the previous models. It was found that the models **Photo_Hinv1**and **Photo_Hinv2a** do not function properly in the acceleration mode in the versions R2016b and R2019a. Therefore, the models **Photo_Hinv1N_2016** and **Photo_Hinv2aN** are developed instead of the formers for these versions. The **Full-Bridge MMC** block is used in the models **Photo_Hinv1NN** and **Photo_Hinv2aNN** for R2019a version.

3.4 PV Island Operation

In the above-considered models, PVs operate in parallel with the power grid that supplies load under bad illumination. However, the situations often occur when PVs are located on the isolated area where the power and reliable electrical grid is absent. Because the production of the electrical energy by PVs has a volatile character, the other energy sources are necessary to secure an uninterrupted electrical supply of the consumers in this area. In this way, the hybrid electrical grids or microgrids are formed. Simulation of such systems is considered in Chapter 6.

In some cases, this isolated area has a connection with the outer electrical network, but this tie is low-powered or (and) unreliable, and can be disconnected at any point; at this, the PV set that connected with it comes to the islanding operation mode. The PV control system has to identify such conditions and to turn off from the grid in order to avoid possible unwanted occurrences that can appear during the following uncontrolled grid voltage emergence. Besides, the certain actions have to be made in the PV set, for example, load reduction or its complete cutoff. Therefore, the PV control system must have a possibility to reveal the islanding mode. Besides, the measures have to be provided for smooth connection to the grid on its recovery.

There are passive and active methods to identify an islanding mode. In the former case, the conclusion is made by observing variations of the quantities during normal operation: variations of an amplitude, a frequency, and waveforms of the load voltage and current. With these methods, the islanding mode can be determined with certainty not in all cases. If, for instance, the power produced by PV is equal to the load power, the grid current is not used, and it can be disconnected without influencing the process in the load. Though, the opportunity remains to monitor the content of the higher harmonics in the load voltage. The point is that the PV inverter fabricates the voltage with some content of the higher harmonics. For these harmonics, the grid reactance is much less than the load resistance that is shunted with the grid; and the higher harmonics do not manifest themselves in the load voltage. When the grid disconnects, the presence of these higher harmonics can be detected in the voltage and in this way the islanding mode is determined. This method is not always reliable, because content of the higher harmonics depends on many factors and THD change is not always large enough.

The certain disturbances are introduced in the system in the active methods, to which it reacts in different way under connected or disconnected grid.

Such systems are single-phase and often have a small power. In the model **Photo_m01** the load power is 10 kW at a voltage of ~110 V. The power circuits are depicted in Figure 3.17. The contactless switches **Br** and **Br1** connect the load for supply from the grid and PV, respectively. **Br2** opening imitates the grid voltage dropping. The PV control system contains the controller of the PV output voltage, whose reference is the PV optimal voltage value that is determined by MPPT; the controller output is the reference of the inverter current I^*. This current is in phase with the grid voltage when **Br** is on or with the waveform $\sin(2\pi f + \varphi)$, where f is the grid frequency. When PV operates separately, the angle φ is not defined (in fact, it is fixed at the moment of the **Br** last disconnection). Before beginning a joint operation (before **Br** closing) the signal PBr is formed; at this, the angle φ is getting equal to the grid phase angle $\angle U_{gr} = \alpha$; so, after **Br** closing, a voltage sudden change is absent.

The error under angle φ control is determined as $\varepsilon = \sin(\alpha - \varphi)$; but because $\varepsilon = 0$ not only under $\alpha - \varphi = 0$ but also under $\alpha - \varphi = \pi$, the quantity $\varepsilon = \sin \alpha \cos \varphi - \cos \alpha \sin \varphi - (1 - \cos \alpha \cos \varphi - \sin \alpha \sin \varphi)$ is utilized, because the bracketed expression is zero at $\alpha - \varphi = 0$, but not zero at $\alpha - \varphi = \pi$.

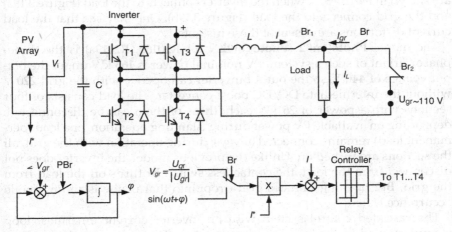

FIGURE 3.17
Single-phase PV set in the model **Photo_m01**.

The following process is simulated. **Br2** closes initially. Immediately after the process beginning, the trigger sets in the subsystem **Switching Control**; 0.8 s later **Br** closes, load supply carries out by the grid. At that, the synchronization of the angle φ with the grid phase was fulfilled. **Br1** closes at $t = 1.2$ s; the inverter current exceeds the load current; a part of the inverter current delivers to the grid.

Br2 opens at $t = 2$ s. Unit **Isl** detects this fact (see further) and gives the signal to reset the trigger and to open **Br**, the PV set comes in the islanding mode. Br2 closes at $t = 3$ s again. When the voltage U_{gr} at the input terminals of **Br** reaches 90 V the trigger sets, the preliminary signal PBr is fabricated that initiates the synchronization of the angle φ with the grid phase. **BR** closes 0.8 s later; PV begins to operate in parallel with the grid.

Two variants to discover an islanding mode are provided in the model. When **Tumbler1** is in low position and **Tumbler2** in top position, this mode is detected by the THD value in the voltage at the input terminals **Br**. In the given case, THD increases from ~0 to 0.02 when coming in the islanding mode. When the tumblers are in the opposite positions, the waveform $10 \times \sin (4 \times 2\pi f)$ is added to the signal I; as a result, the third and fifth harmonics having amplitudes about 2.7% and 2.1%, respectively, appear in the islanding mode; these harmonics are picked up by the Fourier processors. When the sum of the harmonics reaches the given value (5 A in the case), the signal to disconnect **Br** is fabricated.

Some simulation results are depicted in Figures 3.18 and 3.19. The fabrication of the signals is shown in Figure 3.18, which indicate origination of islanding mode (Figure 3.18a, b without introduction of the additional perturbations, Figure 3.18c, d with these perturbations). It is seen that at $t = 2$ s, when the grid is disconnected, these signals appear that results **Br** disconnection at $t = 2.16$ s. The currents of the grid, of the load, and of the inverter are shown in Figure 3.19, when the inverter connects to the load (Figure 3.19a) and the grid connects to the load (Figure 3.19b). One may see that the load current distortions are absent at these instants.

The model **Photo_m02** (**Photo_m02N** for R2016b, R2019a) is the three-phase variant of such a system. PV nominal power is 165 kW under the output voltage of 440 V, so the output inverter can operate with the grid 220 V without the intermediate DC/DC boost converter. The load consists of four sections with a power of 25 kW each; three of them can be disconnected, depending on available PV power during islanding operation, one load (permanent load) remains connected always; during operation with the grid, all the sections are turned on. Unlike the previous model, the inverter does not disconnect from the load, the contactless switch **Br** turns off the load from the grid, **Br2** opening imitates grid dropping, that is, the islanding mode occurrence.

The cascaded control system with an inverter current innermost loop is implemented in the inverter control system. Depending on the mode of operation, the outer loop is either PV voltage loop when PV operates jointly

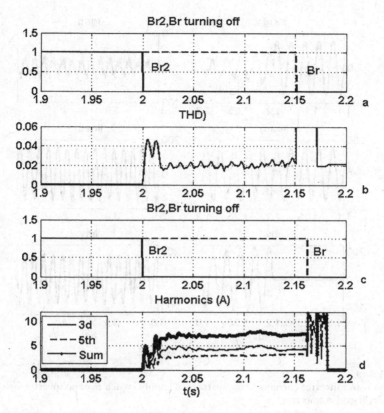

FIGURE 3.18
Identification of the island mode in the model **Photo_m01** (a, b) passive method; and (c, d) active method.

with the grid or the load voltage loop in the islanding mode. In the first case, the reference is the PV optimal voltage under actual condition of environment. To determine the islanding mode, the three-phase waveform of small amplitude is added to the current reference. The waveform frequency is 250 Hz. The block **Three-Phase Sine Generator** fabricates the current reference three-phase waveform; the block output is modulated in amplitude by outputs of PV voltage control or the load voltage control, as it was mentioned above. When **Br** is closed the frequency and the phase of the output waveform is determined by the frequency and phase of the block **PLL**, whose input is the grid voltage, they are constant in the islanding mode. On giving command to turn on **Br**, that is, to work together with the grid, the preliminary signal PBr is formed; with this signal, the phase of **Three-Phase Sine Generator** is adjusted in accordance with the grid voltage phase, and only thereafter, **Br** is closed. **Br** is turned off, when the islanding mode is revealed; in the considered case, when the relative value of sum of three amplitudes of

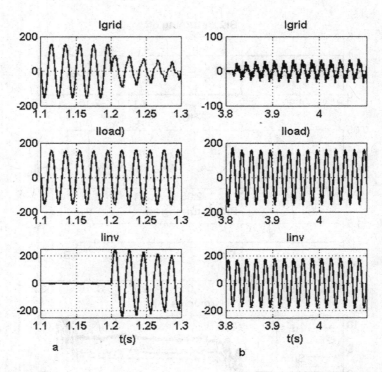

FIGURE 3.19
Grid, load, and inverter currents at the instants of supply switch over (a) inverter switches to
the load; (b) grid connects.

the fifth harmonic voltages at the input **Br** terminals reaches 0.03. The command to turn on **Br** is given when, in the islanding mode, the voltage at its input reaches the value close to the nominal. The model schematic diagram is depicted in Figure 3.20.

PV voltage V_{pv} is not controlled in the islanding mode; it is supposed that the circuit parameters are taken in the way that it always exceeds the optimal value V_{pvopt} so that PV operates on the right branch of the curve $P = f(V_{pv})$ that corresponds to stable conditions: with an increase in load, the voltage V_{pv} decreases, PV power rises, a new stability can be reached. In the islanding mode, it is necessary to reduce load power with the decrease of the power generated by PV; otherwise, its output voltage can drop greatly, and a collapse of the DC bus voltage of the PV can come. The decrease of the difference $V_{pv} - V_{pvopt}$ can be an indicator of the necessity to disconnect a part of the load, but such a system is difficult to implement in practice, because the given difference can reduce fast, and making a decision about which parts of the load must be deenergized in given conditions, demands some time, during which the collapse of the DC bus voltage can occur.

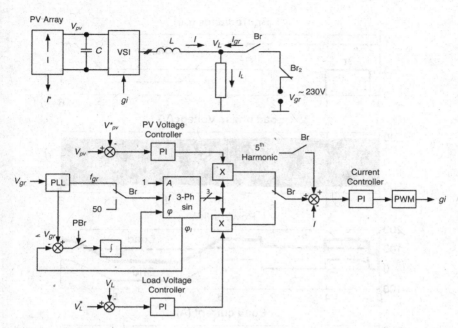

FIGURE 3.20
Schematic diagram of the model **Photo_m02**.

Therefore, it is reasonable to assume that, in order to switch load parts promptly, the possibility must be to determine the maximal obtainable load power P_m in present conditions. The voltage optimal value is determined in any case for system operation. The current optimal value can be determined with help of the test PV that runs under the short-circuit conditions, taking into account the factor K_i (Section 3.1). The product of the voltage and current optimal quantities determines the maximal obtainable power. The losses can be taken into account when the powers are calculated, at which the connections/disconnections of the separate load parts take place. Repeated simulation of the considered system under different load powers gives the ratio of the load power to the PV power as 0.915; that is, PV power 27.3 kW gives a load power of 25 kW. The switching points are chosen in accordance with given relationship in the subsystem **Load**.

The results of simulation of the following sequence of the events are displayed in Figure 3.21: joint operation with the grid under the nominal irradiance; grid fall out at $t = 1$ s; its recovery at $t = 2$ s, irradiance reduction to 0.25 at $t = 3$ s, and its increase to 1 at $t = 5$ s. The load power keeps invariable. When the value of irradiance is nominal, the part of PV power injects in the grid. When the irradiance drops, the load is supplied from the grid, in which the power transfer direction changes. When the grid is disconnected, the load power is provided by PV. Grid connection to the load goes smoothly.

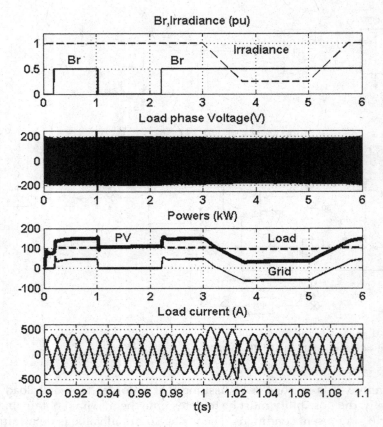

FIGURE 3.21
Processes in the model **Photo_m02** during joint and separate operations with the grid.

A short-term increase of the load current with duration of 1 grid voltage period can be seen, because the current controller needs some time to suppress the excess current that had been injected in the grid before.

The operation in islanding mode is shown in Figure 3.22, when the irradiance reduces from 1 to 0.25. One may see, as the irradiance decreases, the PV power reduces, which results in the successive disconnection of the load parts and decrease of the power. The load voltage is kept constant.

Instead of two test PVs, one can be employed, with switching over that puts it in short-circuits and no-load conditions by turns. Such the system is realized in the model **Photo_m02a (Photo_m02aN** for R2016b, R2019a). **Ideal Switch** turns on and turns off in turns, by control of the pulse waveform. If this waveform is equal to 1, short-circuits condition arises. Before this waveform takes zero value, the PV output current is kept in the block

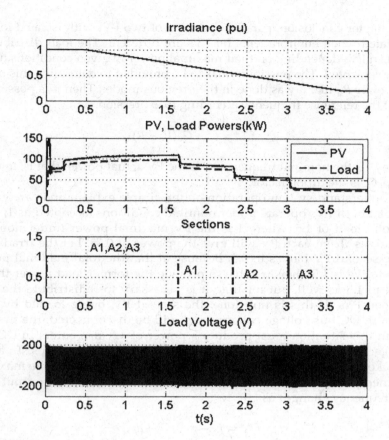

FIGURE 3.22
Processes in the model **Photo_m02** under irradiance drop.

Sample and Hold. If the waveform is equal to zero, no-load condition has place. Before the waveform takes value 1, the PV output voltage is kept in another block **Sample and Hold**. The outputs of the both blocks are multiplied, so the resultant output is the maximal power. This output is delayed by one period of the pulse waveform that, under abrupt change of environment, can lead to collapse the DC bus voltage. In actual fact, environmental variations move relatively slow. The impulse period is taken as 0.25 s in the considered model, in which processes do not differ noticeably from those shown for model **Photo_m02**.

The model PV from Ref. [3] is employed in the models **Photo_m02_2015a** and **Photo_m02a_2015a** (**Photo_m02N_2015a** and **Photo_m02aN_2015a**, respectively, for R2016b, R2019a); PV parameters are close to those used in the previous models. The processes in the models differ slightly.

In chapter conclusion, parallel operation of two PVs with isolated load is simulated when environments for PVs are different. The load distribution between PVs depending on their maximal power in given conditions has to be implemented. The possibility has to be provided to determine this maximal power P_m like it was done in the previous model. Then it is possible to adjust the reference frequency according to expression

$$f^{**} = f^* + G_p(P_m - P) \tag{3.6}$$

where f^* is the nominal frequency, P is the PV actual power, that is, to use a droop, G_p is the droop factor.

To understand system operation better, simple explanations are given. First, let both PVs operate under nominal conditions (accepted as 1) with the total load of 1.5, where 1 is the PV maximal power under nominal conditions. Then each PV will give the power of 0.75. Let the irradiance of the second PV halves to 0.5. Because at that the total maximal power of both PVs is 1.5, they are able to provide the former load power that is equal to 1.5 as well, but for that, it is necessary to redistribute the load between PVs. If this is not done, the second PV, being loaded by 0.75, enters in DC bus voltage collapse and will be disconnected; the first PV that in this situation must produce a power of 1.5 also enters in DC bus voltage collapse, and production of the electrical energy stops. To avoid a blackout, the PV control system must sense a decrease of the maximal obtained power and react to it. In the model **Photo_m03** the output voltage frequency changes as

$$f^{**} = f^* - g_p P/P_m + df, \tag{3.7}$$

which is a modification of Equation (3.6) under $G_p = g_p/P_m$, df is an additional correcting signal of the secondary loop. It is accepted $g_p = 1$.

The same PVs as in the previous model are utilized. The reactors at the inverter outputs are connected in parallel to the common load. The AC voltage controllers without innermost current controllers are used in the inverter control systems.

The output of the voltage controller modulates in amplitude three-phase waveform of **Sine Generator** with the controlled frequency. Two possible load circuits are provided. When **Tumbler 2** is in the position L, **Load** is active; in position L1, **Load1** is active. Accordingly, the order of irradiation variation change.

Let us carry out simulation with **Tumbler 2** in the position L. The load 150 kW increases to 220 kW at $t = 0.5$ s. The irradiance of PV1 is 1, the irradiance PV2 drops to 0.5 at $t = 1$ s. With the invariable frequency $f^{**} = f^* = 50$ Hz, the load voltage reduces from 200 V to 150 V and the load power drops from 220 kW to 135 kW after PV2 irradiance decreases, that is the

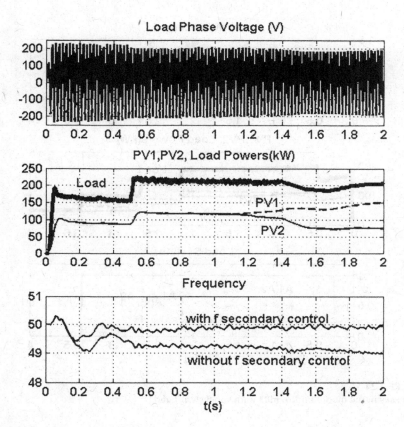

FIGURE 3.23
Processes in the model **Photo_m03** with subsystem **Load**.

normal work stops. With droop utilization by Equation (3.7), the load voltage will be 190 V and load power 210 kW; the PV powers are 150 kW and 75 kW, respectively, but the frequency reduces to about 49 Hz. With employment of the frequency secondary loop (**Tumbler 1** is closed), the frequency recovers (Figure 3.23).

The load **Load1** consists of four sections, each of them has a power of 50 kW. They are switched over, depending on a sum of the maximal powers of both PVs. The process is shown in Figure 3.24 with **Tumbler 2** in position L1 in the case when the irradiance of PV1 drops to the value of 0.25 with the rate of 0.25 1/s, beginning at $t = 0.5$ s; the irradiance of PV2 reduces also to the value of 0.25 with the same rate, but beginning at $t = 1$ s. With the radiance decrease, the power of both PVs drops that results in the successive disconnection of the load sections and decrease its power. The load voltage is kept constant.

The model PV from Ref. [3] is used in the model **Photo_m03_2015a**.

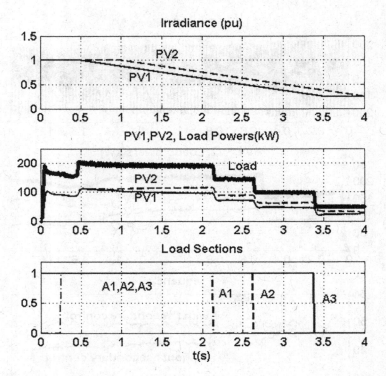

FIGURE 3.24
Processes in the model **Photo_m03** with subsystem **Load1**.

References

1. Perelmuter, V. Renewable Energy Systems, Simulation with Simulink® and SimPowerSystems. CRC Press, Boca Raton, FL, 2017.
2. Sera, D., Mathe, L., Kerekes, T., Spataru, S. V., and Teodorescu, R. On the Perturb-and-observe and incremental conductance MPPT methods for PV systems. IEEE Journal of Photovoltaics, 3(3), 2013, July, 1070–1078.
3. MathWorks, Simscape™ Electrical™, User's Guide (Specialized Power Systems). MathWorks, Natick, MA, 1998–2019.
4. Blaabjerg, F., and Ionel, D. M. (ed.) Renewable Energy Devices and Systems with Simulations in MATLAB® and ANSYS®. CRC Press, Boca Raton, FL, 2017.
5. Mirhosseini, M., Pou, J., and Agelidis, V. G. Single- and two-stage inverter-based grid-connected photovoltaic power plants with ride-through capability under grid faults. IEEE Transactions on Sustainable Energy, 6(3), 2015, July, 1150–1159.

6. Xiao, B., Hang, L., Riley, C., Tolbert, L. M., and Ozpineci, B. Three-phase modular cascaded H-Bridge multilevel inverter with individual MPPT for grid-connected photovoltaic systems. Twenty-Eighth Annual IEEE Applied Power Electronics Conference and Exposition (APEC), 17–21 March, 2013, Long Beach, CA, 468–474.
7. Rivera, S., Kouro, S., Wu, B., Leon, J. I., Rodríguez, J., and Franquelo, L. G. Cascaded H-Bridge multilevel converter multistring topology for large scale photovoltaic systems. IEEE International Symposium on Industrial Electronics, 2011, 1837–1844.
8. Sochor, P., and Akagi, H., Theoretical and experimental comparison between phase-shifted PWM and level-shifted PWM in a modular multilevel SDBC inverter for utility-scale photovoltaic applications. IEEE Transactions on Industry Applications, 53(5), 2017, September/October, 4695–4707.
9. Fuentes, C. D., Rojas, C. A., Renaudineau, H., Kouro, S., Perez, M. A., and Meynard, T. Experimental validation of a single DC bus cascaded H-bridge multilevel inverter for multistring photovoltaic systems. IEEE Transactions on Industrial Electronics, 64(2), 2017, February, 930–934.

4

Fuel Cell and Microturbine Simulation

4.1 Simulation of the Fuel Cells and Electrolyzers

The fuel cell (FC) is a device that directly converts fuel chemical energy into electric energy. During this process, the intermediate conversion of the chemical energy into the thermal or the mechanical energy does not take place, hence the efficiency of the device can be very high. Unlike the battery, which must be disconnected periodically for recharge, the FC can produce electric energy without interruption, as long as it contains the fuel (such as hydrogen) and an oxidant (such as oxygen), which can be refilled during FC operation.

There are various types of FC. The most widespread are the polymer electrolyte fuel cells (PEMFCs) and the solid oxide fuel cells (SOFCs). The formers are the low-temperature devices and operate at temperatures around 80°C; pure hydrogen is necessary for their function that can be obtained either with the help of a reformer from a natural gas or by an electrolysis. They have rather small start time from a cold state, so they can be utilized as back-up units in microgrids, and also in vehicles.

The SOFC is a high-temperature device, whose working temperature can reach 1000°C. So, it can have the internal reformer and, therefore, employ, as a fuel, the natural gas directly. In FC, water and heat are generated and released as by-products that must be removed and can be used effectively. Besides, some amount of gaseous fuel remains disused and appears at the FC outlet. Hot gas present at the SOFC outlet has high temperature, so the systems are worked out, in which SOFCs operate together with the microturbines (MTs); they, in the process, produce the hot gas as well. Utilization of the hot gas exiting from one device to heat the gases entering the other device improves the total efficiency of the whole set, Ref. [1].

FC sets are rather complicated systems that have the units for fuel, oxidant, water supply, for withdrawal of the exhaust gas and the generated water, and for temperature maintenance. However, when the FC set is investigated as the additional source in the systems with wind generator (WG) or Photovoltaic (PV), a number of simplifications are taken. It is often supposed that the controllers work ideally, and the only channel for action on the output voltage is the hydrogen consumption.

PEMFC is considered at first. There is a general formula for the main part of the FC voltage, Ref. [2]

$$E = E^0 + \frac{RT}{2F} \ln\left(\frac{P_{H_2}\sqrt{P_{O_2}}}{P_{H_2O}}\right), \qquad (4.1)$$

where E^0 is the open circuit voltage, R is the universal gas constant that is equal to 8314 J/kmol/K, T is the absolute temperature (K), F is the Faraday's constant equal to 96.485×10^6 C/kmol, and P_{H_2}, P_{O_2}, P_{H_2O} are the partial pressures of hydrogen, oxygen, and water vapor, respectively (on the cathode). Water does not evaporate at the low temperature of PEMFC, so the term P_{H_2O} in Equation (4.1) can be omitted. Using the data from Ref. [2] and making computation, Equation (4.1) can be written as

$$E = E_0^0 - 8.48 \times 10^{-4} \times (T - 298.15) + 4.31 \times 10^{-5} T \ln\left(P_{H_2}\sqrt{P_{O_2}}\right), \qquad (4.2)$$

where E_0^0 is the open circuit voltage under standard pressure and temperature and with the pure reagents; it will be taken as $E_0^0 = 1.229$ V.

When the FC conducts the current, the voltage drop appears, so it may be written as

$$V = E - ir - \Delta V_{act} - \Delta V_{con}, \qquad (4.3)$$

where i is the FC current density, mA/cm²; r is the resistance of the whole electrical circuit including the membrane and various interconnections, $k\Omega$ cm²; ΔV_{act} is the activation losses, owing to the slowness of the polarization reactions taking place on the electrodes; and ΔV_{con} is the voltage drop, owing to the change of the concentration of the active components in the reaction area.

The appropriate expressions are intricate and depend on many parameters, so that for the creation of the FC model, the empirical and semiempirical formulas are used, as, for instance

$$\Delta V_{act} = A \ln\left(\frac{i}{i_0}\right), \qquad (4.4)$$

where i is the FC current density, mA/cm²; i_0 is the density of some reference current; A is the some constant; the formula is valid under $i > i_0$.

If in Equation (4.3), instead of E we substitute $E_C = E + A \ln(i_0)$, then

$$\Delta V_{act} = A \ln(i). \qquad (4.5)$$

The following formula is suggested for ΔV_{con} in Ref. [2]

$$\Delta V_{con} = me^{ni}, \tag{4.6}$$

where m and n are constants.

Data for PEMFC ($T = 70°C$) are given as, Ref. [2], follows: $i_0 = 0.04$ (mA/cm^2), $r = 2.45 \times 10^{-4}$ (kΩ cm^2), $A = 0.03$ (V), $m = 2.11 \times 10^{-5}$ (V), $n = 8 \times 10^{-3}$ (cm^2/mA). The ratio of oxygen/hydrogen consumption R_{HO} is often given. The equations for the partial pressures of hydrogen and oxygen are written as

$$\tau_h \frac{dP_{H_2}}{dt} + P_{H_2} = \frac{1}{K_{H_2}}(q_{H_2} - 2K_r I) \tag{4.7}$$

$$\tau_o \frac{dP_{O_2}}{dt} + P_{O_2} = \frac{1}{K_{O_2}}(q_{H_2}/R_{HO} - K_r I) \tag{4.8}$$

$$K_r = \frac{N}{4F}. \tag{4.9}$$

Here, N is the number of series connected cells; q_{H_2} is the inlet flow rate of hydrogen, kmol/s; K_{H_2} and K_{O_2} are the valve molar constants for hydrogen and oxygen, respectively; τ_h and τ_o are the response times for hydrogen and oxygen, respectively; and I is FC current (A).

The FC model contains a model of the gas feed mechanism that consists of PI controller and the second-order unit, Figure 4.1. This unit can model either the reformer that converts a natural gas into the pure hydrogen or a driver of the valve that adjusts the rate of the hydrogen consumption.

The circuits are provided in the FC model, which enable to simulate an operation break approximately. The control signal *Contr* = 0 when FC is working. When this signal comes over in the state 1, the valve that controls hydrogen supply begins to close. When ingress of hydrogen drops essentially, the signal *F_Stop* is generated that decreases a flow rate of all components to zero. The limiting circuits are included in the model, so that it

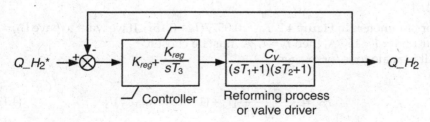

FIGURE 4.1
Block diagram of model of the gas feed mechanism.

can continue to work with integrator zero's outputs. The zero of the output voltage is obtained at the expense of the negative value of the second item in Equation (4.1). Because the multiplier before logarithm is small, the absolute value of the logarithm must be large, that is, the algorithm argument has to be very small on the order of 10^{-40}. A large amount of time is necessary to reach this small value, for instance, 150–200 s. Such simulation duration can be inadmissibly large for simulation together with the other devices and components of power electronics. To decrease the FC voltage decay time, **Switch2** is placed in the FC model that determines, at which value of the quantity, $P_{H_2}\sqrt{P_{O_2}}$, the minimal value of the algorithm argument is given. In order to have current smooth rise at fuel feed start, the first-order filter with a time constant of 3 s is placed in the current circuit, as in the model in Ref. [3].

When the control signal *Contr* gets to zero, the hydrogen supply renews, the pressures and the voltage at output of the block **Controlled Voltage source** recover.

The described model of FC is utilized in the model **Fuel_1**. The FC has a nominal voltage of 400 V and a current of 150 A. Its dialog box is displayed in Figure 4.2. The value of fuel rate has two components: the constant one and the second that is proportional to FC current,

$$q_{H_2} = q_{H_2}^0 + 2K_r I/K_f, \tag{4.10}$$

where K_f is the fuel utilization factor, $q_{H_2}^0$ is proportional to some value of the initial current I_0:

$$q_{H_2}^0 = 2K_r I_0/K_f. \tag{4.11}$$

Then, using Equations (4.7), (4.10), and (4.11), the P_{H_2} value in the steady state can be obtained as

$$P_{H_2} = \frac{2K_r}{K_{H_2}K_f}[I_0 + (1 - K_f)I]. \tag{4.12}$$

For parameters in Figure 4.2, $P_{H_2} = 0.0577(I_0 + 0.15I)$. If we want to have $P_{H_2} = 4$ bar under $I = 150$ A, then $I_0 = 47$ A must be chosen.

It is analogous for oxygen

$$P_{O_2} = \frac{2K_r}{K_{O_2}K_f R_{HO}}[I_0 + (1 - 0.5 \times R_{HO}K_f)I]. \tag{4.13}$$

$P_{O_2} = 3.1$ bar under $I = 150$ A.

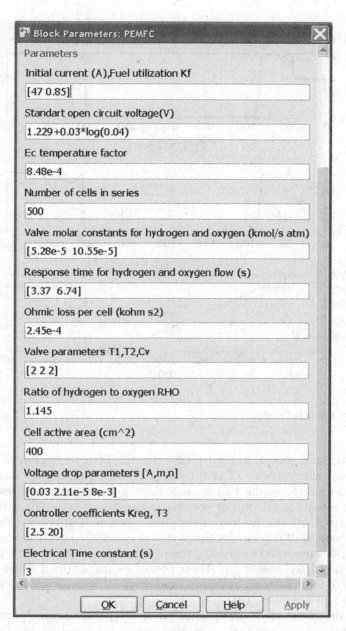

FIGURE 4.2
Dialog box of the developed FC.

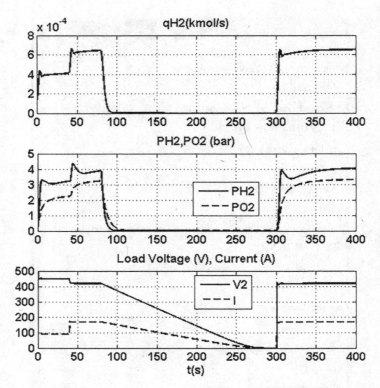

FIGURE 4.3
Processes in the model **Fuel_1**.

The process is shown in Figure 4.3 when the FC load resistance that was 5 Ω initially decreases to 2.5 Ω at $t = 40$ s. When signal *Contr* = 1 at $t = 80$ s, hydrogen feed stops, the hydrogen and oxygen partial pressures drop to zero that results in FC output voltage decreasing to zero. When the signal *Contr* = 0 at $t = 300$ s, hydrogen supply renews, and afterward the output voltage recovers. It is seen that the partial pressures in steady state are equal to the above-computed quantities; the processes of a stop and the following start are simulated adequately. These processes are obtained when the point *Threshold* of **Switch2** is set as zero. To set *Threshold* = 3×10^{-11}, the output voltage would be zero at $t = 150$ s.

These situations often occur when there are possible short pauses in FC operations due to some fault or by fuel supply interruption. But load uninterrupted supply has to be provided, therefore, a device for the electrical energy storing is necessary. A supercapacitor (SC) is used for this aim in the model **Fuel2**. It is connected to FC output via a diode and an insulated gate bipolar transistor (IGBT) conducting current in the opposite directions, dispensing with the DC/DC converter, Figure 4.4. The fuel rate is a sum of two

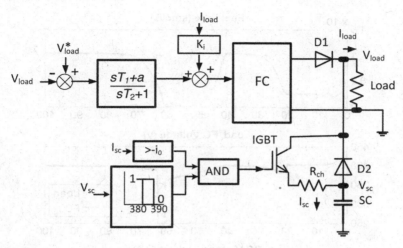

FIGURE 4.4
Block diagram of the set with FC and SC.

components: one of them is proportional to FC current and another is the output of the controller of FC voltage, the voltage reference is 400 V. When the FC voltage drops lower than the SC voltage, the diode begins to conduct, and the SC determines the load voltage and discharges. When the FC voltage increases, the diode is cut off, and the SC is charged through the IGBT. The charge stops when the SC voltage increases to the required value.

The SCs BCAP2000 are used during simulation having parameters 2000 F, 2.7 V, $R_s = 0.35$ mΩ, 110 A. The SC unit consists of two parallel circuits; each of them has 160 capacitors connected in series. In the considered model, the fuel feed stops at $t = 40$ s and renews at $t = 60$ s. The process is shown in Figure 4.5. One can see that the system copes with failure, whereas without SC circuits the normal operation is not possible. The duration of the complete disappearance of FC voltage is 4 s. At this time, the load voltage and its current (about 250 A) are determined by the SC voltage solely that decreases by $\Delta U \approx 250 \times 4/(2000/160 \times 2) = 40$ V. After FC feed renewal, its voltage recovers and SC voltage rises.

Such a system can be employed for the load voltage drop decrease under load abrupt increase, considering FC slow response for external action.

The systems in which natural gas is used to obtain hydrogen cannot be considered as fully renewable ones, because natural gas does not recover. Situation changes when hydrogen comes out from water by electrolysis. Because the electrolysis demands an electrical energy, the latter is taken from renewable energy systems (RESs) in the time when its production exceeds the load needs. The produced hydrogen is stored in the tank either under high pressure or in liquid form and is used for FC supply at the times, when RES powers are deficient.

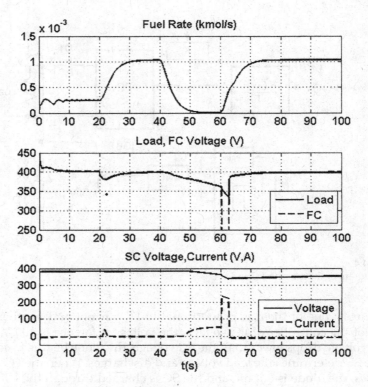

FIGURE 4.5
Processes in the model **Fuel2** with supercapacitor.

The electrolyzer is a device that, substantially, is opposite to FC: if, in the latter, hydrogen and oxygen turn into the electrical energy and water, the water, with the help of electricity, turns into hydrogen and oxygen in the former. There are alkaline and proton exchange membrane electrolyzers.

The rate of hydrogen production is defined by Faraday's Law

$$\dot{n}_{H_2} = \eta_F \frac{n_c I}{2F},$$
(4.14)

where I is the electrolyzer current (A), F is the Faraday's constant (96,487 C/mol), n_c is the number of electrolyzer cells in series, η_F is Faraday's efficiency factor that is computed by the different empirical formulas, for example, Refs. [4, 5]:

$$\eta_F = 0.96 \frac{\left(\dfrac{I}{A_F}\right)^2}{\left(\dfrac{I}{A_F}\right)^2 + 2.5 \times 10^4},$$
(4.15)

where A_F is the area of cell (m^2).

For supplying DC voltage, the electrolyzer is a nonlinear resistance, whose characteristic can be taken as, Ref. [5]

$$U = n_c U_c,$$ (4.16)

$$U_c = U_r + \frac{r}{A_F} I + s \ln\left(\frac{t}{A_F} I + 1\right).$$ (4.17)

Parameters U_r, r, s, and t depend on temperature T (°C) and on electrolyzer type. They are found by statistical treatment of the experimental data. The heating process is not simulated in the following models; some mean dependence for the alkaline electrolyzer with the cell area of 0.25 m^2 and the rated current 750 A is accepted that was calculated with the use of data given in Refs. [4, 5]

$$U_c = 1.22 + 6.44 \times 10^{-5} \times I + 0.185 \times \ln(1 + 0.09 I).$$ (4.18)

This dependence is plotted in Figure 4.6. If to take $n_c = 21$, then, with the voltage 21 × 2.06 = 43.26 V, the power is 32 kW and productivity

$$\dot{n}_{H_2} = 0.96 \frac{9 \times 10^6}{9 \times 10^6 + 2.5 \times 10^4} \times \frac{21 \times 750}{2 \times 96487} = 0.0783 \text{ mol/s} = 282 \text{ mol/h}$$

which gives for ideal gas under temperature of 25°C, $\dot{n}_{H_2} = 282 \times 0.0245 = 6.9$ m^3/h.

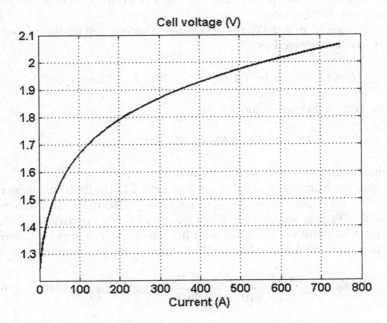

FIGURE 4.6
Dependence $U_c = f(I)$ for alkaline electrolyzer.

Investigations are activated lately as for proton exchange membrane electrolyzers. PEM-based electrolysis has many advantages when compared to conventional alkaline-based electrolysis, for example, it has smaller dimension and mass, lower power consumption, intrinsic ability to cope with transient electrical power variations, a high degree of purity of gases, and the potential to compress hydrogen at a higher pressure within the unit and with higher safety level.

There are rather many works in which the models of such electrolyzers are considered, for instance, Ref. [6], but they are rather complicated and need a number of parameters that are not given always in electrolyzer technical data. In order to simulate microgrid with FC and electrolyzer, it is sufficient to have the dependence of the electrolyzer current on the applied voltage, as the rate of the hydrogen production is determined by Equation (4.14) as before. It is supposed that the electrolyzer is supplied from the controlled DC current source. It follows from data given in the references that the considered curve $U_c = f(i)$, $i = I/A_F$ has a short steep, linear or nonlinear, at $i < 0.1$ A/sm^2 approximately and the practically linear dependence under further current density increase. These curves depend on the temperature, so the needed voltage value decreases with its rise, but taking into account that consideration of this factor requires knowledge of the thermal model of the device, some of the average characteristics are used further. It is reasonable to have in mind that, under current control, inaccurate knowledge of the dependence considered affects, to some extent, the transient, but not the hydrogen production rate.

From data given in Ref. [7], the dependence for one cell under temperature of $\approx 40°$C can be received as

$$U_c = 1.3 + 3i \text{ under } i \leq 0.1; U_c = 0.78i + 1.522 \text{ under } i > 0.1. \qquad (4.19)$$

Or as an analytical expression

$$U_C = 1.123 + 0.65i + 0.05 \times \ln(1 + 10800i). \qquad (4.20)$$

The relevant curves are plotted in Figure 4.7.

One may see from Figure 4.3 that, under load 2.5 Ω, the hydrogen flow rate is 0.63 (mol/s), that is, the hydrogen required is $0.63 \times 3600 \times 0.0245 = 55.6$ m^3/h $= 55.6 \times 0.09 = 5$ kg/h, where 0.09 kg is a weight of 1 m^3 of hydrogen. Let's suppose that the alkaline electrolyzer with 105 cells and a cell area of 0.25 m^2 and rated current of 750 A is employed. Then, electrolyzer productivity will be

$$\dot{n}_{H_2} = 0.96 \frac{9 \times 10^6}{9 \times 10^6 + 2.5 \times 10^4} \times \frac{105 \times 750}{2 \times 96487} = 0.392 \text{ mol/s} = 1410 \text{ mol/h}$$

which makes for ideal gas under temperature 25°C, $\dot{n}_{H_2} = 1410 \times 0.0245 = 34.5$ m^3/h.

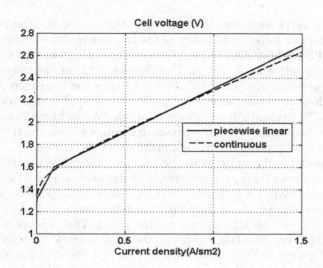

FIGURE 4.7
Dependence of the voltage from current density for PEM electrolyzer.

Suppose that the hydrogen store has to provide the operation with the rated load within 8 h; it means that the hydrogen store under nominal conditions has to be $55.6 \times 8 = 445$ m³. Suppose that the hydrogen is kept in the tank under pressure of 40 bar under 30°C (303 K). Then the tank volume has to be about (see formula (4.21) below with $a = b = 0$)

$$V = (445/0.0245) \times 303 \times 8.314/40/10^5 = 11.4 \text{ m}^3.$$

The tank is filled up for $445/34.5 \approx 13$ h.

The more precise formula for pressure in the tank, Ref. [8], is used for subsequent simulation:

$$P = \frac{nRT}{V - nb} - \frac{an^2}{V^2} \tag{4.21}$$

$$a = \frac{27R^2T_{cr}^2}{64P_{cr}}, \qquad b = \frac{RT_{cr}}{8P_{cr}}, \tag{4.22}$$

where
  P is the pressure, Pa;
  n is the number of moles, mol;
  R is the universal gas constant, 8.314 J/K/mol;
  T is the temperature, K;
  V is the volume of storage tank, m³;
  T_{cr} is the critical temperature, 33.2 K for hydrogen; and
  P_{cr} is the critical pressure, 1.28×10^6 Pa for hydrogen.

The model **electr1** simulates the joint operation of the FC, the electrolyzer and the hydrogen tank. Availability of the DC source 400 V is assumed. The subsystem **PS** contains DC/DC buck converter with an output current controller, whose reference is 750 A. The characteristic of electrolyzer having 105 cells is given above. The block **Fcn** calculates the hydrogen production rate. To reduce the simulation time considerably, this rate is increased by 1000 times. The pressure in the tank is computed in the block **Pressure** using above-given formulas. When the pressure reaches 40 bar, the trigger resets, tank filling up stops, and operating permit for FC is formed.

The FC model is the same as in the previous model. When FC is operating, the contents of the hydrogen tank and the pressure in it decrease, and the rate of the hydrogen consumption increases 1000 times by simulation as well. When the pressure drops to 1% of the maximal value, FC operation stops, and the tank filling up begins again. The process is shown in Figure 4.8. The real time taken to fill up the tank is 45 s × 1000/3600 = 12.5 h, FC operating time is 30 s × 1000/3600 = 8.3 h that fits with the above-given calculations (13 h and 8 h).

These models are utilized in Chapter 6 for hybrid system simulation.

The SOFC model is considered further. These FCs have the big starting time so they are not fit for utilization in the often start-stop modes. Therefore, the appropriate circuits are removed from the SOFC model to be present in

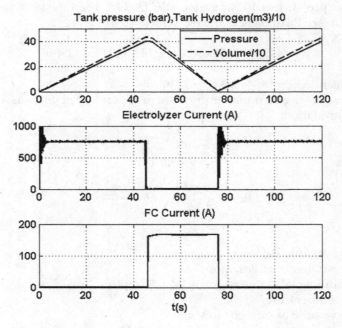

FIGURE 4.8
Simulation of joint operation of FC, the electrolyzer, and the hydrogen tank.

the PEMFC model. As T_n is the designated nominal temperature for which the parameters given in the dialog box are defined; T_t is the current temperature. At that, the heating process is not simulated because it goes on dozens minutes; instead, processes at any temperature can be simulated.

For SOFC, the main part of the FC voltage is determined in Equation (4.1), so, in addition to Equations (4.7) and (4.8), the equation for water vapor has to be added

$$\tau_{H_2O} \frac{dP_{H_2O}}{dt} + P_{H_2O} = \frac{2K_r I}{K_{H_2O}}. \tag{4.23}$$

The time constants τ_x in Equations (4.7), (4.8), and (4.23) are inversely proportional to the temperature value, Ref. [4], so they are calculated as $\tau_x = \tau_{xn} T_n / T_t$, where τ_{xn} is the value of the respective time constant in the dialog box.

Considering the data given in Ref. [2], instead of Equation (4.2), Equation (4.24) is used

$$E = E_0^0 - 0.3 \times 10^{-3} \times (T - 1273) + 4.31 \times 10^{-5} T \ln\left(P_{H_2} \sqrt{P_{O_2}} / P_{H_2O}\right), \tag{4.24}$$

where the open circuit voltage under standard pressure and temperature is taken as $E_0^0 = 1$ V.

The actual temperature is taken into account under computation of the activation and ohmic losses. According to Ref. [4], they decrease when the temperature increases. Here the linear dependences are accepted:

$$\Delta V_{act} = A \ln (i)\left[1 - K_a (T_t - T_n)\right] \tag{4.25}$$

$$r = r_0 \left[1 - K_b (T_t - T_n)\right]. \tag{4.26}$$

The values A and r_0 are taken as 0.02 V and 2.45×10^{-4} (kΩ cm^2), respectively, the K_a and K_b coefficients approximately correspond to the plots given in Ref. [4]: loss variation by 25–30% under temperature variation by 100°C.

The model **Fuel_sof** is intended for determination of the static and dynamic characteristics of the FC model developed. Its power is 60 kW (400 V, 150 A). The model dialog box is shown in Figure 4.9. If to fix $R_1 = R_2 = 2.5$ Ω, Br opening time as 50 s, and to disconnect Br1, the process shown in Figure 4.10 is obtained that describes system dynamics.

The resistor R_3 is the smoothly time-varying resistance borrowed from the model, **Photo_m1**. If to fix very small resistance R_1, for example, $R_1 = 2.5 \times 10^{-4}$ Ω, very large resistance R_2, for example, $R_2 = 2.5 \times 10^4$ Ω, to open Br and to close Br1, FC $V - I$ dependence under given temperature can be received after carrying out simulation. These dependencies are plotted in Figure 4.11 for 800°C, 900°C, and 1000°C. It should be noted that these characteristics are obtained with the current feedback, when, with the current decrease, the fuel consumption and the partial pressure decrease too.

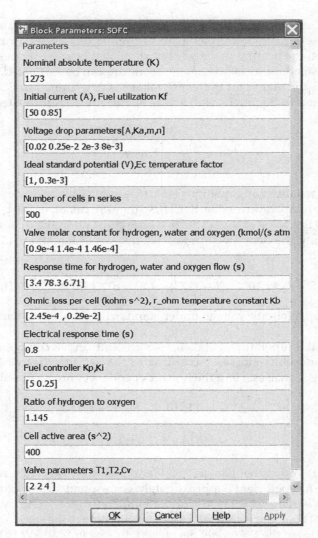

FIGURE 4.9
SOFC model dialog box.

In the model, the relationships (4.24) and (4.26) are realized by two components: the main part with help of the DC voltage source block and resistor block, the part depending on temperature with utilization of the Simulink® blocks, and the block **Controlled Voltage Source**. The parameters accepted for the subsystem **Valve** are usual for the reformer, Ref. [1].

In the model **Fuel_sof1**, FC, through voltage source inverters (VSI), supplies the load of 30 kW at the voltage of 230 V, the power surplus is sent to the AC grid. The FC from the previous model is employed. The VSI control

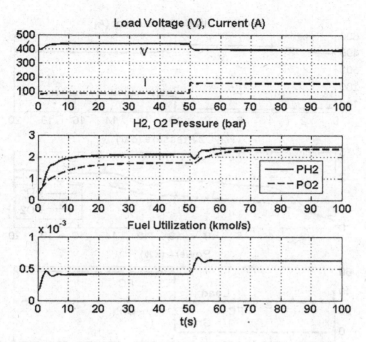

FIGURE 4.10
Transients in the SOFC model.

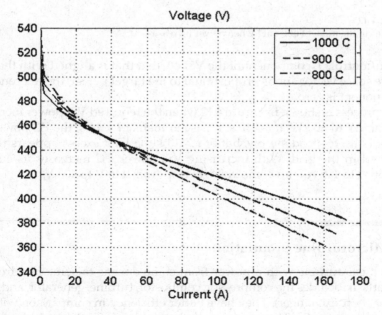

FIGURE 4.11
SOFC characteristics under different temperatures.

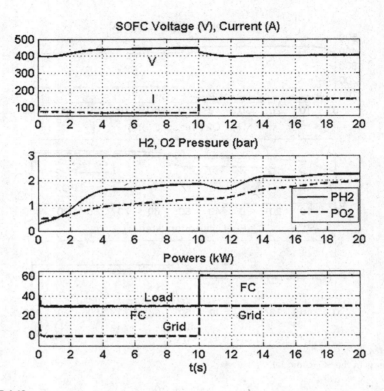

FIGURE 4.12
Processes in the model **Fuel_sof1** when power increases.

system controls I_d component of the VSI current that is aligned with the grid voltage space vector and is proportional to the target power; the reference for I_q component is zero.

The process is shown in Figure 4.12 when the required VSI power increases from 30 kW to 60 kW at $t = 10$ s. It is seen that, at $t < 10$ s, the FC power are used wholly for load supply, but at $t > 10$ s the load excess appears that is delivered in the grid. With increasing VSI power, FC increases its current from 66 A to 149 A with some voltage drop from 446 V to 403 V.

4.2 Microturbine Simulation

MT is a gas turbine with a power from some dozens to some hundreds of kilowatts. It includes the compressor, combustor, turbine, generator, and recuperator (heat exchanger). They have better efficiency in comparison with the diesel generator sets. The pressurized air, together with the fuel, comes into the combustion chamber where the mixture is burned. The gas of the high pressure and the high temperature that is fabricated in so doing rotates the turbine, which, in turn, rotates the compressor and generator. With help of recuperator, the hot exhaust gas of the turbine (typically around 900 K) pre-heats the compressed air (typically around 420 K) going into the combustor.

There are the single-shaft and the split-shaft MTs. In the former, the turbine, the compressor, and the generator, which are mounted on the same shaft, have very large rotation speed—up to 100,000 r/min. Therefore, the generator has a very high output frequency and requires additional units of the industrial electronics. In the latter, the generator is driven via a gearbox, so that the generator can have the standard frequency of the output voltage and dispense with additional devices for connecting with the grid. The former becomes widespread today because it has less rotating parts. The PMSG with one pole pair is used in the modern types of the MTs, so that the output voltage frequency can reach 1500–1600 Hz. The rotation speed and the torque of the turbine are controlled with the rate of the fuel supply.

The control system has to ensure not only the system function under normal operation conditions, but also the specific modes of starting and stopping. Before fuel injection and ignition, the turbine must be speeded up to a preset rotational speed. The external source of the energy is necessary; as such, the grid or the battery can be used; the generator in the motor mode can be utilized for speeding up. Therefore, the electronic devices at the generator output have to fulfill bidirectional power transfer or an additional inverter can be used.

One can distinguish between cooldown, or normal, and warmdown stop sequences. In the former, the fuel supply is first cut off; the generator begins to decelerate, producing some amount of electric energy. When the thermal energy stored in the recuperator and the other elements of the system decreases to such a low level that the compressor stops operation, the generator passes into the motor mode, supporting the airflow through the system for its cooling, until its temperature reaches the preset level.

At this point, the turbine may be quickly stopped by simply shorting the generator terminals.

Under the warmdown stop, the kinetic energy of the rotating parts is absorbed by the brake resistor, as in the dynamic braking mode of the electrical drive, Ref. [9].

A number of MT models exist, which, with different accuracy, take into account the thermodynamic processes in them, Ref. [10]. For MT modeling in electric power systems, more simple models are of interest, in which the start and stop processes are not considered. The most popular model developed in Ref. [11] is shown in Figure 4.13.

The speed controller affects the rate of the fuel supply V_c and has the transfer function as

$$F_c = \frac{K(sT_1 + 1)}{sT_2 + a}. \tag{4.27}$$

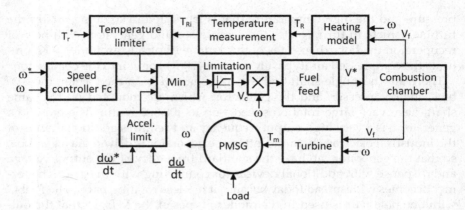

FIGURE 4.13
Block diagram of the microturbine model.

The rate of the fuel supply may be decreased if the turbine acceleration exceeds the permissible limit (that can have a place, e.g., under abrupt load release) and if the temperature of the exhaust gases exceeds the limit. The acceleration limiter is an integrator; the temperature limiter is a PI controller. The least of the three signals determines the value V_c.

The fuel supply system is described by the relationship

$$V^* = (V_{c0} + K_1 \omega V_c)\frac{1}{sT_3 + 1}\frac{1}{sT_4 + 1}, \qquad (4.28)$$

where V_{c0} is the rate fuel supply in the no-load condition, the first transfer function of the first order defines the fuel valve lag, the second one, the actuator lag. It is taken in pu $V_{c0} = 0.23$, $K_1 = 0.77$. The process in the combustion chamber (its output is V_f) is modeled with the delay θ_C. The torque developed by the turbine may be computed as

$$T_m = 1.3\left(\frac{1}{sT_5 + 1}V_f - 0.23\right) + 0.5(1 - \omega), \qquad (4.29)$$

where T_5 is determined by the gas turbine dynamic. The rotational speed may be found from the Equation (1.17) which is computed in the model of the coupling PMSG; the moment of inertia has to include both the generator inertia and the inertia of the rotating turbine parts.

The exhausting gas temperature (°C) can be found from the relationship

$$T_R = T_R^* - 371(1 - e^{-\theta_e}V_f) + 288(1 - \omega), \qquad (4.30)$$

where θ_e is the exhausting gas reaction delay. The transfer function of the temperature measurer is

$$F_{izm} = \frac{1}{sT_6 + 1}\left(K_2 + \frac{K_3}{sT_7 + 1} \right), \tag{4.31}$$

where T_6 is the time constant of the thermocouple, K_2, K_3, and T_7 are the parameters of the radiation shield ($K_2 + K_3 = 1$), $T_{Ri} = F_{izm}T_R$.

Simulation of the systems with the MT demands a very small sampling period (~1 μs), bearing in mind the high frequency of the generator output voltage, whereas the transients in the turbine run slowly and last minutes or dozens of minutes so that the simulation time turns out to be very large. Therefore, a number of simplified models are developed in Ref. [12], which, with the reasonable duration of simulation, give the opportunity to study and investigate the system key features. Furthermore, the more complicated models are considered.

In the model **MTURBO1M**, MT has a power of 60 kW and the rotational speed of $2400 \times \pi$ rad/s, that is, the nominal torque is 8 Nm. With a rotor flux of 0.065 Wb, the PMSG rms line voltage without load is 600 V. The turbine model carries out the relationships, Equations (4.27) to (4.31); the Simulink blocks of the transfer functions, which enable to specify the initial outputs are used under model building that gives an opportunity to increase slightly a simulation speed. The turbine fuel control system keeps the turbine nominal rotational speed with a slight increase of 8% under the torque rise from 0 to its nominal value.

PMSG is linked to the grid with the help of two VSI (VSI-Ge and VSI-Gr) connected back-to-back with the common DC link. The reference of the DC voltage is set as 900 V. VSI-Ge control system assigns the power that has to be taken from PMSG adjusting the current I_d component that is aligned with the PMSG voltage, Figure 4.14. VSI-Gr is connected with the 380-V grid and with the load of 50 kW power. Its control system keeps DC voltage adjusting the current I_d component that is aligned with the grid voltage.

The process is simulated when the reference power increases from 40 kW to 60 kW at $t = 2$ s, Figure 4.15. It is seen that system response is rather rapid. First, the grid takes part in the load supply, but at $t = 2$ s, power excess appears, and power surplus goes to the grid.

In the model **MTURBO2M**, the MT supplies an isolated load. The VSI-Gr control system changes: it determines the amplitude and the frequency of the load voltage. The DC link voltage is kept by VSI-Ge. The load that was initially equal to 40 kW increases to 60 kW at $t = 2$ s. The variations of the power, DC link voltage, load voltage, and the PMSG current are displayed in Figure 4.16. One can see that two last-named quantities are practically sinusoidal.

In the model **MTURBO1M**, the controllers of the turbine, VSI-Ge, and VSI-Gr have the following functions, respectively: the turbine rotational speed control, required power control, and DC link voltage keeping. But the

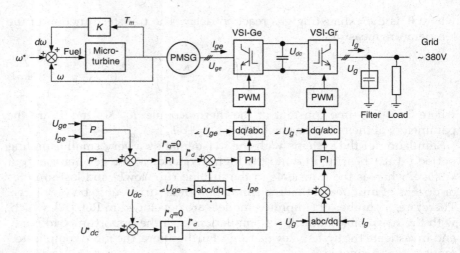

FIGURE 4.14
Block diagram of the system with the microturbine and AC grid.

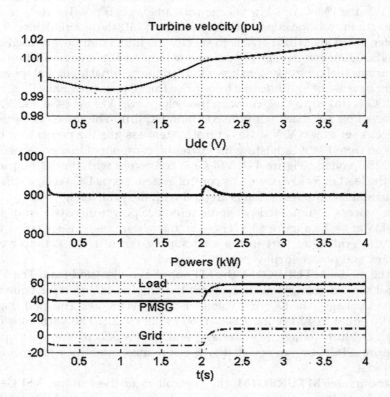

FIGURE 4.15
Processes in the model **MTURBO1M**.

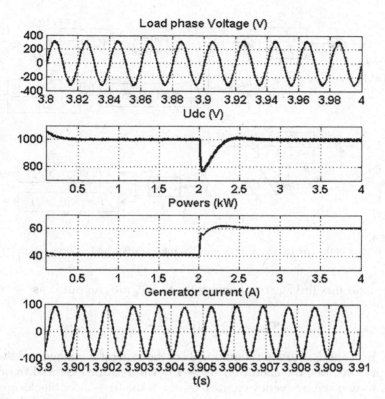

FIGURE 4.16
Processes in the model **MTURBO2M**.

other assignment of the control functions is possible as well. For instance, it is possible that the turbine rotational speed control assigns to VSI-Ge. Such a structure is used when the turbine starts with the utilization of PMSG as a starting engine. Besides, this speed control is faster than impact on the fuel rate; the power is determined by the fuel rate in the case.

The current controllers I_d and I_q are used in the rotational speed control system; these currents are aligned with the rotor position. The output of the speed controller determines the reference for I_q; the reference for I_d is zero. The quantities $k\omega L_s I_q$ and $-k\omega L_s I_d$ are added to the outputs of the controllers I_d and I_q, respectively, where L_s is the inductance of the stator winding, and k is the compensation factor.

Because the measurement of the generator rotational speed and rotor position is not a simple task, taking in mind the large first quantity, the operation without the mechanical transducers is considered, the so-called, sensorless control. The block diagram of the circuits for the estimation of the rotor position and its rotating speed are shown in Figure 4.17.

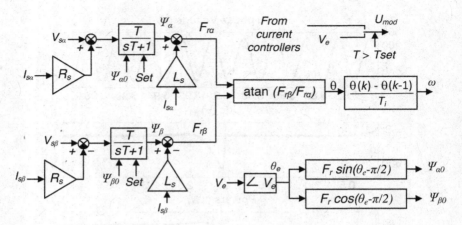

FIGURE 4.17
Block diagram of the circuits for estimation of the rotor position and rotating speed.

The stator flux linkage components Ψ_α and Ψ_β are computed as

$$\Psi_{\alpha(\beta)} = \frac{T}{sT+1}(V_{s\alpha(\beta)} - R_s I_{s\alpha(\beta)}),\qquad(4.32)$$

where $V_{s\alpha(\beta)}$ and $I_{s\alpha(\beta)}$ are the components of the generator voltage and the current space vectors; the condition $T\omega \gg 1$ is valid in the essential frequency range, so that the frequency characteristics of the first-order blocks are very close to the integrator frequency characteristics; and R_s is the stator winding resistance. The components of the rotor flux that are created with the permanent magnets are

$$F_{r\alpha(\beta)} = \Psi_{\alpha(\beta)} - L_s I_{s\alpha(\beta)}.\qquad(4.33)$$

The rotor position may be found as

$$\theta = arctg\left(\frac{F_{r\beta}}{F_{r\alpha}}\right),\qquad(4.34)$$

and the rotational speed as

$$\omega(k) = \frac{\theta(k) - \theta(k-1) + 2\pi a}{T_i},\qquad(4.35)$$

where T_i is the sampling period, $a = 1$, if $\theta(k)$ and $\theta(k-1)$ are taken from the different periods of rotor rotation (from the different teeth of the sawtooth waveform, $\theta = f(t)$), $a = 0$ in the rest of the cases.

The described system is implemented in the model **MTURBO3M**. The power circuits are the same as in the model **MTURBO1M**. MT power control is carried out: the reference comes at input P^*, the feedback $P = T_m\omega$ comes at input P. To speed up simulation, the acceleration limitation is increased from 0.05 to 0.5.

VSI-Ge control system keeps the turbine rotational speed equal to 7500 rad/s, having the innermost controllers of the currents I_d, I_q. The estimations of the rotational speed and rotor position are carried out with relationships in Equations (4.32) to (4.35). Pay attention that, to estimate flux linkages by Equation (4.32), the specially developed blocks are utilized, enabling designation of the required initial values when the input pulse signal is active; this feature is used in the next model **MTURBO4M**. The VSI-Gr control system is the same as in the model **MTURBO1M**, the DC link voltage is increased to 1100 V.

The process is shown in Figure 4.18 when the MT wants to increase power from $P^* = 0.7$ to 0.9 at $t = 4$ s. The initial segment is not simulated. It is seen that system response is quiet adequate, the rotational speed and DC link voltage are kept constant.

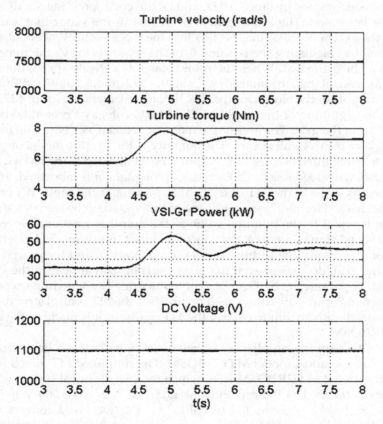

FIGURE 4.18
Processes in the model **MTURBO3M**.

MT start is simulated in the models **MTURBO4M** and **MTURBO4M1**. The difference is that, in the latter, the full model of the MT is employed, as in previous models, whereas its simplified version is used in the former: only the controller transfer function block retains, and also the torque computation by Equation (4.29) is carried out: all the transfer functions and the delay units are excluded. The simulation in this model runs about twice quicker, but the transients of powers and speed are the same practically.

The structure of these models is the same as in the model **MTURBO3M**. The speed reference increases from 150 rad/s to 7500 rad/s with an acceleration of 500 rad/s^2. Measurement of the rotational speed and rotor position is carried out with the same system as in the previous model. The difference is in the circuits for estimation of the rotor flux linkage initial values.

In order for the system to work properly, the initial estimation of the flux linkage has to be completed under the start that is according to the real rotor position. It is carried out in the following way. During the start, after the DC link voltage had reached the steady state, the modulation voltage V_e of the small frequency, for instance, 1 Hz, and of the corresponding small amplitude is fabricated. The rotating field is created in the stator that leads the rotor; the axis of the rotor magnet flux lags the space vector V_e at about $\pi/2$. In this way, by calculating the position θ_e of the space vector V_e, the rotor position may be estimated. When the rotor rotates at a steady speed, the signal *Set* is formed; at that, the quantities $F_r \sin(\theta_e - \pi/2)$ and $F_r \cos(\theta_e - \pi/2)$ are set at the outputs of the blocks for the flux linkage estimation, Figure 4.17.

At the beginning of the speed up, the PMSG runs as a motor, taking the power from the grid. The antitorque moment is taken as 1 Nm. The input of the turbine control unit is zero. The quantity at the turbine model output T_m may be disregarded because this output is not connected to the PMSG. When the rotational speed reaches 5000 rad/s, the signal *On* is fabricated, and the fuel begins to go to the MT (the signal appearing at the input *W'*) producing the driving torque. The magnitude of this signal is chosen such that the torque is equal to the nominal value at the nominal rotational speed; the actual speed signal is zero at the controller input in this mode. Under influence of this torque, the turbine keeps on accelerating, and the power, taken from the grid, decreases; even the power surplus can appear in the system with other parameters that can be sent to the grid. The speeding-up process is shown in Figure 4.19. After speed up finishes, the MT control structure can be changed with an aim to control the MT power as it is made in the model **MTURBO3M**.

When the turbine supplies an isolated load, a battery can be utilized for turbine acceleration, model **MTURBO5M**. The simplified MT model is used as in the model **MTURBO4M** from which the speed control system without sensors is taken. The battery with the nominal voltage of 300 V is placed in the DC link. It is connected to VSI-Ge via DC/DC boost converter with the output voltage of 900 V. During acceleration, VSI-Gr (VSI-Load in the case) is disconnected from the DC link. When the rotational speed reaches

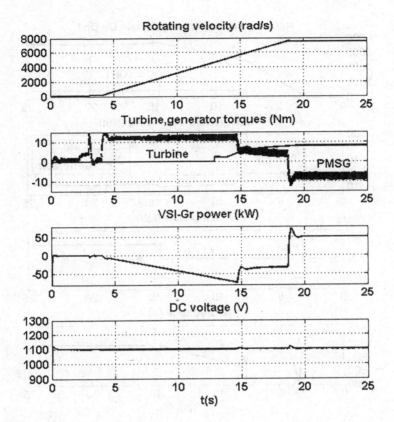

FIGURE 4.19
Microturbine speeding up in the model **MTURBO4M**.

5000 rad/s, the MT is activated and produces the additional torque, operating in speed control mode for preparation for the autonomous work. When the rotational speed comes nearer to the nominal one, the signal (named *On*1) is fabricated; the battery is disconnected from the DC link and VSI-Load, on the contrary, is connected to the DC link, fabricating the required load voltage 380 V, 50 Hz. Load power is 40 kW. VSI-Ge control system switches over to the DC link voltage control, as in the model **MTURBO2M**, with a reference value of 1000 V. The process is shown in Figure 4.20. It is seen that the speed up ends at about $t = 15$ s, the load 41 kW is switched to the MT. When VSI-Load is connected, DC voltage is kept constant.

Along with the above-investigated system with two back-to-back VSIs (and even more often, Ref. [13]), the system with an input diode rectifier is used, Figure 4.21. Such the system allows dispensing with the power transistors switching over with the high frequency but initiates a number of other problems. Because the rectifier transfers energy only in one direction, either a mechanical starter or the backup low-power inverter is needed; the latter can

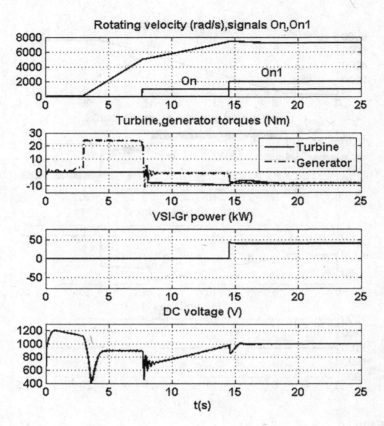

FIGURE 4.20
Processes in the model **MTURBO5M**.

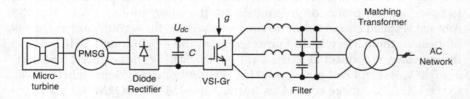

FIGURE 4.21
Block diagram of the system with the MT and diode rectifier.

be supplied from the grid (if available) or from the battery. Besides, such a system has lesser flexibility with the parameter selection because the rectifier output voltage is determined uniquely by EMF of PMSG and the voltage drops across its inductance and the filter inductance and decreases essentially with the increase in load.

The above disadvantage can be eliminated with the employment of the DC/DC boost converter between the diode rectifier and VSI-Gr, but the turbine start problem remains. These complicated systems are not considered further.

The structure shown in Figure 4.21 is simulated in the model **MTURBO6M**. The MT model is the same as in the previous models but, in order to gain the power of 60 kW, it turned out to be necessary to increase the PMSG rotor flux linkage by 10%, that is, to use a greater generator. The MT control system keeps its rotational speed equal to the nominal one, with the slight rise when the PMSG torque increases. VSI-Gr determines the power that is sent to the grid and local load by controlling I_d current component that is aligned with the inverter output voltage space vector. The grid and load voltages, and the load power are the same as in the model **MTURBO1M**, that is, 380 V, 50 kW. The DC voltage in this system is much less than in the mentioned model, therefore, connection with the grid is carried out with the help of the transformer 220/380 V. The process is shown in Figure 4.22 when the

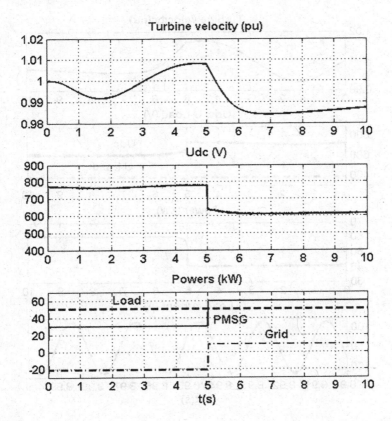

FIGURE 4.22
Processes in the model **MTURBO6M**.

required power increases from 30 kW to 60 kW at $t = 5$ s. It is seen that at the beginning, the grid takes part in the load supply, but at $t = 5$ s power excess appears, and power surplus goes to the grid.

The model **MTURBO61M** simulates the similar system but with an autonomous operation as in the model **MTURBO2M**. The load inverter determines the amplitude and frequency of the load voltage. The process is displayed in Figure 4.23 when the load rises from 40 kW to 60 kW at $t = 5$ s. It is seen that the load voltage is constant and the current is nearly sinusoidal.

The MTs can be utilized together with the other sources of the electric energy making the hybrid systems. For example, a number of articles deal with the combined use of MTs and high-temperature FCs. Although both of these devices have the same ability to produce electricity continuously, regardless of climatic conditions, their certain connection by exchanging the outgoing gases can significantly increase their overall efficiency (see below). Of particular interest are the hybrid schemes MT-WG, Ref. [14], and MT-FC, Ref. [15].

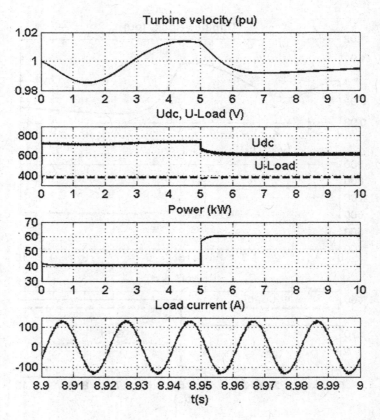

FIGURE 4.23
Processes in the model **MTURBO61M**.

The operation of the turbine has some peculiarities in comparison with the diesel generator. Starting a turbine from a cold state takes several minutes and requires an external source of energy and increased fuel consumption. In addition, turbine exhaust gases can be used to heat water or room (the so-called combined heat and power [CHP] systems), so it can be assumed that in many hybrid systems the MT will function continuously. Then the role of RES is to reduce the MT fuel consumption and to power additional loads if there is sufficient power.

In the model **MTURBO7M**, the joint operation of the MT and the WG is simulated. The first one is made according to the model **MTURBO2M** scheme, its power is tripled to 180 kW. Recall that the turbine control system regulates its rotational speed, the VSI-Ge control system maintains the voltage in the DC link, and the VSI-Gr generates a voltage of 380 V at a frequency of 50 Hz in the microgrid. WG power is 250 kW at a wind speed of 12 m/s. Its scheme is the same as in the model **Wind_PMSG_9N**, the output inverter control system maintains the voltage in the DC link, ensuring the transfer of generated power to the grid.

The load contains a non-disconnected (critical) part with a power of 150 kW and two additional loads with a power of 30 kW each, which are connected when the sufficient wind speed exists. At the low wind speed, the critical load is supplied only from the MT; with an increase in wind speed, the WG begins to take part in load power, while the turbine fuel consumption decreases. With the further increase in wind speed, additional loads are connected in series. Since, at certain wind speeds, the power generated by WG may exceed the load power, the turbine generator will go into motor mode, which seems undesirable. To prevent this, it is necessary either to reduce the power of the WG in some way, or to use a ballast load. In this model, the ballast load is applied, which is activated when the power of the MT is reduced to 0.05 nominal, preventing the generator from going into engine mode.

In the model under consideration, the initial wind speed is 7.8 m/s, which corresponds to a power of 69 kW, so that the powers of the turbine and the wind turbine are about the same, giving a total of 150 kW. At $t = 3$ s, the wind speed begins to increase to 12 m/s with an intensity of 1 m/s^2. The power change process is shown in Figure 4.24. It can be seen that the power of the MT at some moments decreases to zero, but does not become negative; one can be sure that without a ballast load this fact would have occurred. Figure 4.25 shows that the load voltage and its frequency are kept with a high degree of accuracy.

Although the MT operates continuously, the WG can shut off at light wind or during repairs, so it is advisable to simulate the process of putting the WG into operation. This model is presented in the model **MTURBO7M1**. Initially it is assumed that the wind speed is zero (or that the WG is braked or that its blades are turned away from the wind). Load supply is provided by the MT. In this case, VSI-Gr of WG is connected to the microgrid, and the voltage

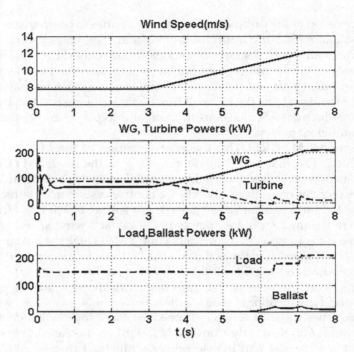

FIGURE 4.24
Processes in the model **MTURBO7M**.

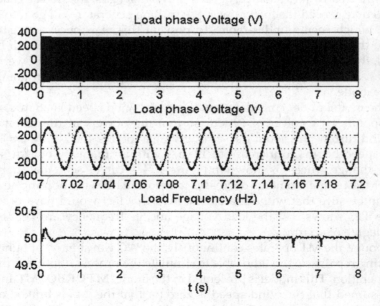

FIGURE 4.25
Load voltage and frequency in the model **MTURBO7M**.

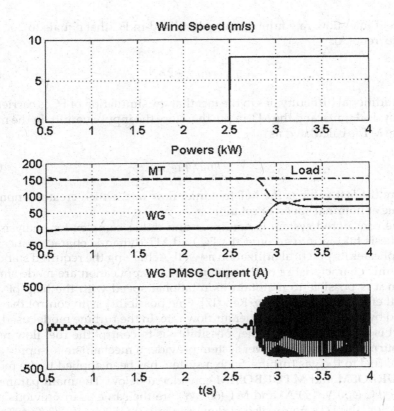

FIGURE 4.26
Processes in the model **MTURBO7M1** under wind speed increase.

in the WG DC link is maintained equal to the specified one. At $t = 2.5$ s, the wind turbine is driven by the wind with a speed of 7.8 m/s, it accelerates, the generator current and the turbine moment increase, and it takes upon itself a part of the load, as was the case in the previous model. The process is shown in Figure 4.26.

We have already mentioned above the feasibility of the joint operation of the MT and high-temperature FC using the mutual exchange of outgoing gases. With the most popular layout of the system, the hot gas emitted from the MT heats the fuel and the air at the inlet of the FC, and the rest of the fuel is used as the fuel for the turbine. Thus, in this case, FC and MT are an integrated structure with a common control system. It is often assumed that in such a hybrid system the SOFC channel will supply approximately 80% of the total output power and the MT channel will supply 20%, Ref. [16].

The full model of such a system requires the use of thermodynamics and electrochemistry ratios, which is not envisaged in the scope of this book, so that simplified models will be used. It is assumed that the temperature of the FC is somehow optimized and assigned. MT fuel inlet flow rate is equal to

FC fuel inlet flow rate minus the fuel utilized in FC that equals to $2 K_r I$, I is FC current (A):

$$q_{H_2t} = q_{H_2f} - 2K_r I. \tag{4.36}$$

An additional difficulty lies in the fact that the simulation of FC is carried out in physical units, and the MT in pu, therefore the input quantity in the model of the MT is calculated as

$$V_c = q_{H_2t} / q_{H_2t\_nom}, \tag{4.37}$$

where the last value is the MT fuel inlet flow rate, which creates a nominal torque at nominal speed of rotation.

The considered system is nonlinear and rather complex for analysis and synthesis taking into account the FC and MT dynamic characteristics; this complicates its practical utilization as well. Achieving the required static and dynamic characteristics of the system and its application are made simpler when it is possible to maintain its rotational speed with the help of additional effects on the turbine, Ref. [17]. One possibility is to control the shaft speed by controlling the burner air flow rate. In the turbine model used, this effect is not provided. Another possibility is to control the fuel flow rate to the burner if the hybrid power system provides a mechanism to supply additional fuel to the gas turbine. Such a system has been applied to the models **MTURBO8M** and **MTURBO81M** considered below. The main parameters of the FC (400 V, 150 A) and MT (60 kW) are the same as in previous models. Thus, the FC's nominal fuel flow rate is $150 \times 500 \times 10^{-3}/2/96487 \approx 4 \times 10^{-4}$ kmol/s. It is assumed that the nominal flow rate is equal to the same value: $q_{H_2t\_nom} = 4 \times 10^{-4}$ kmol/s. MT fuel flow rate is

$$V_c = (q_{H_2f} - 2K_r I + \Delta q_{H_2})/q_{H_2t\_nom}, \tag{4.38}$$

where Δq_{H_2} is the output of the turbine rotational speed controller.

MT generator is connected to the system output via a diode rectifier and a DC/DC booster converter. In the same way, FC is connected to the system output, that is, both DC/DC converters are connected in parallel. Their output voltages are regulated with the introduction of droop terms, so that references for the output voltages for the FC and MT are equal, respectively

$$U_f^* = U_{c0} - D_f I_f, \quad U_t^* = U_{c0} - D_t I_t,$$

where D_f and D_t are the droop factors, I_f and I_t are output currents of the respective DC/DC converters, $U_{c0} = 1000$ V. Because it is supposed that the usable capacity of FC surpasses the same of MT, it is accepted $D_t > D_f$, Figure 4.27.

In the model **MTURBO8M**, DC load that was equal to 50 kW increases to 80 kW at $t = 10$ s, $D_t = 4D_f$. The process is displayed in Figure 4.28.

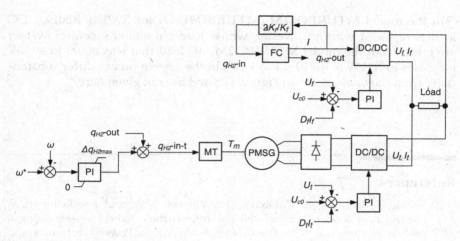

FIGURE 4.27
Block diagram of the FC and microturbine parallel operation.

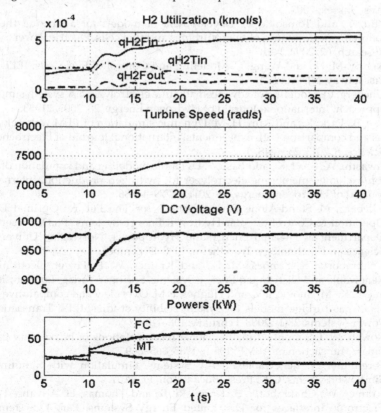

FIGURE 4.28
Processes in the model **MTURBO8M**.

In the model **MTURBO81M** (**MTURBO81MN** for R2016b, R2019a), DC load is replaced with an inverter, whose load circuits and control system are taken from the model **MTURBO2M**. AC load that was equal to 40 kW increases to 70 kW at $t = 10$ s. Processes in the system do not differ substantially from the ones shown in Figure 4.28 and are not given here.

References

1. Vechiu, I., Baudoin, S., Camblong, H., Vinassa, J.-M., and Kreckelbergh, S. Control of a solid oxide fuel cell/gas microturbine hybrid system using a multilevel convertor. 17th European Conference on Power Electronics and Applications (EPE'15), Geneva, Switzerland, 2015, 1–8.
2. Larminie, J., and Dicks, A. Fuel Cell Systems Explained. John Wiley & Sons Ltd, West Sussex, 2003.
3. Zhu, Y., and Tomsovic, K. Development of models for analyzing the load-following performance of microturbines and fuel cells. Electric Power Systems Research, 62, 2002, 1–11.
4. Nehrir, M. H., and Wang, C. Modeling and Control of Fuel Cells. IEEE Press, Piscataway, NJ, 2009.
5. Ulleberg, Ø. Modeling of advanced alkaline electrolyzers: A system simulation approach. International Journal of Hydrogen Energy, 28, 2003, 21–33.
6. Lee, B., Park, K., and Kim, H.-M. Dynamic simulation of PEM water electrolysis and comparison with experiments. International Journal of Electrochemical Science, 8, 2013, 235–248.
7. Awasthi, A., Scott, K., and Basu, S. Dynamic modeling and simulation of a proton exchange membrane electrolyzer for hydrogen production. International Journal of Hydrogen Energy, 36, 2011, 14779–14786.
8. Ulleberg, Ø. Stand-Alone Power Systems for The Future: Optimal Design, Operation & Control of Solar-Hydrogen Energy Systems. Ph.D. Dissertation, Department of Thermal Energy and Hydropower, Norwegian University of Science and Technology, Trondheim, 1998, December.
9. Chakraborty, S., Simões, M. G., and Kramer, W. E. Power Electronics for Renewable and Distributed Energy Systems. Springer-Verlag, London, 2013.
10. Yee, S. K., Milanovic, J. V., and Hughes, F. M. Overview and comparative analysis of gas turbine models for system stability studies. IEEE Transactions on Power Systems, 23(1), 2008, February, 108–118.
11. Rowen, W. I. Simplified mathematical representations of heavy-duty gas turbines. Transaction of ASME, 105(1), 1983, 865–869.
12. Perelmuter, V. Renewable Energy Systems, Simulation with Simulink and SimPowerSystems. CRC Press, Boca Raton, FL, 2017.
13. Kramer, W., Chakraborty, S., Kroposki, B., and Thomas, H. Advanced Power Electronic Interfaces for Distributed Energy Systems, Part 1: Systems and Topologies. Technical Report NREL/TP-581-42672, National Renewable Energy Laboratory, Golden, CO, 2008, March.

14. Jain, A., Singh, B. P., Bhullar, S., and Verma, M. K. Performance of hybrid wind-microturbine generation system in isolated mode. International Conference on Emerging Trends in Electrical, Electronics and Sustainable Energy Systems (ICETEESES-16), 2016, 64–70.

15. Lee, Y.-D., Chang, Y.-R., Chan, C.-M., and Ho, Y.-H. Preliminary implementation of microgrid with photovoltaic and microturbine for stand-alone operation. IEEE Industry Applications Society Annual Meeting, 2012, 1–9.

16. Xu, M., Wang, C., Qiu, Y., Lu, B., Lee, F. C., and Kopasakis, G. Control and simulation for hybrid solid oxide fuel cell power systems. Twenty-First Annual IEEE Applied Power Electronics Conference and Exposition (APEC'06), 2006, 1269–1275.

17. Abbasi, A., and Jiang, Z. Power sharing control of fuel cell/gas-turbine hybrid power systems. IEEE Electric Ship Technologies Symposium, 2009, 581–588.

5

Hydro and Marine Power Plants

5.1 River Hydro Stations

Water covers more than 70% of the earth's surface and possesses huge, practically inexhaustible reserves of energy, which, however, is not always easy to harvest and use. The settings that are most commonly used for this purpose are the following:

1. Large hydroelectric power plants
2. Small hydropower plants such as run-of-the-river
3. Ocean wave energy plants
4. Ocean tide energy plants

This type of installation can also be attributed

5. Pumped hydro storage

The first are very large constructions: the river stream is blocked with a dam that forms a large water reservoir, which is intended for both the increase in the water potential energy and water storage during drought periods. A large area of the land is utilized for the building and the follow-up operation of these electric stations. The power of such electric stations can reach some gigawatts. The block **Hydraulic Turbine and Governor** is included in SimPowerSystems that is intended for simulation of such hydropower plants; a number of examples of this block utilization are given in Refs. [1–3].

The small hydroelectric stations have the power from some kilowatts to some megawatts. When they are designed, one tries to use the natural water level difference; the special reservoirs, when they are built, serve only to increase the pressure of the water column, not to accumulate the water storage; the loss of the ground is minimal. Most of such electric stations are a run-of-the-river type. For such electric stations, the normal river stream does not change essentially; the water reservoir is small or is completely absent. The essential part of the water stream comes to the turbine via the canal or the conduit, and then the water returns to the river. The different types of turbines are used in the small hydroelectric sets, including the propeller turbines, the Francis and Pelton turbines. In the propeller turbine, the water stream comes parallel to the axis, and in the Francis turbines, the water stream comes perpendicular to the axis in the radial direction. The turbine axes can be vertical or horizontal and have a slope or a variable blade position (Kaplan turbines).

Power control of the small turbines is carried out by means of the gate or the guide vanes; for turbines with a power less than 100 kW, the dummy load can be used as well. Often the control system has to maintain the upper water level on a fixed level. The synchronous generators (SGs) or the induction generators (IGs) with the squirrel cage rotor are employed in such turbines; when the turbine operates together with the grid, the turbine rotating speed is constant for the former and nearly constant for the latter. But the employment of the turbine with variable rotational speed can improve the turbine operation range and increase energy conversion efficiency under changeable hydrological conditions perceptibly. So, utilization of permanent magnet synchronous generator (PMSG) together with power electronic components is an attractive option. A number of examples of the small hydroelectric station models with the IG and the SG are given in Ref. [3]. Here some models of such sets with the PMSG are considered.

It is shown experimentally in Ref. [4] that the dependences of the turbine torque T_m on its rotating speed ω, for a propeller turbine, under different water discharge Q, are the straight, practically parallel lines with a negative slope:

$$T_m = T_{m0} - b\omega, \tag{5.1}$$

where T_{m0} depends on the water head H and the water discharge Q; at this, H can be taken often as a constant value. Therefore, the turbine power P has the maximum at $\omega = T_{m0}/2b$ that is equal to $P_{\max} = T_{m0}^2/4b$.

In the model **Hydro_PMSG_1**, the PMSG has the power of 250 kW at the speed of 66 rad/s under nominal discharge $Q = 1$. Then it follows from the above given relationships $b = 2.5 \times 10^5/66^2 = 57.4$ Nms, $T_{m0} = 2 \times 57.4 \times 66 = 7577$ Nm. The PMSG has six pole pairs and the line-to-line rms voltage $V_{rms} = 1.052 \times 66 \times 6 \times \sqrt{3}/\sqrt{2} = 510$ V. Circuits of power electronics consist of two voltage source inverters (VSIs) connected back-to-back, as in many other models; the generator VSI (VSI-Ge) controls the PMSG rotational speed, and the grid VSI (VSI-Gr) transfers the fabricated energy to the grid, keeping the capacitor voltage in the DC link. The reference of this voltage is taken as 1200 V. The VSI-Gr supplies the load with a power of 200 kW, 400 V; the excess power goes to the grid. Because utilization of the speed and position sensors in the small hydro turbine control system is undesirable, a sensorless estimation method of these quantities is used that is taken from the model **Wind_PMSG2_sensorless**, Ref. [3].

The sliding mode observer is used. The vector $\mathbf{\Psi}_r$ has the components $\Psi_{r\alpha} = -F_r \sin\theta$, $\Psi_{r\beta} = F_r \cos\theta$, where F_r is the rotor flux and θ is the rotor position. The space vectors of the stator current $\mathbf{i}_s$ and voltage $\mathbf{u}$ are measured. For $\mathbf{\Psi}_r$ estimation, $\hat{\mathbf{i}}_s$ current estimation is used

$$L\frac{d\hat{\mathbf{i}}_s}{dt} = \mathbf{A}_{11}\mathbf{i}_s + \mathbf{u} - \mathbf{z}, \tag{5.2}$$

$\mathbf{z} = K_1\mathbf{sign}(\mathbf{e}_i)$, $\mathbf{e}_i = \hat{\mathbf{i}}_s - \mathbf{i}_s$, $\mathbf{A}_{11} = a\mathbf{I}$, $a = R_s$, R_s, and L are the stator winding resistance and inductance, and $\mathbf{I}$ is an unit matrix.
If the value K_1 is chosen sufficiently large,

$$\mathbf{z} \rightarrow \omega_r[\Psi_{r\beta}, -\Psi_{r\alpha}]. \tag{5.3}$$

After averaging the obtained values of the vector $\mathbf{z}$ components we find $e_\alpha = \omega_r F_r \cos\theta$, $e_\beta = \omega_r F_r \sin\theta$, so that

$$\theta = arctg\frac{e_\beta}{e_\alpha}. \tag{5.4}$$

The rotational speed may be found as

$$\omega_r(k) = \frac{\theta(k) - \theta(k-1)}{T_c}, \tag{5.5}$$

where T_c is the sampling period.
The reference signal for the speed is proportional to the value Q in pu. Therefore, the discharge Q is supposed to be measurable, direct or indirect. The dependences of the optimal rotational speed and the power obtained at the VSI-Gr output on Q are displayed in Figure 5.1. The process is shown in Figure 5.2 when, under initial discharge $Q = 1$, this quantity decreases to $Q = 0.75$ at $t = 15$ s and to $Q = 0.5$ at $t = 30$ s. It is seen that the values of the speed and power given in the Figure 5.1 are obtained. The simulation of this model runs very slowly. Considering that the DC voltage is maintained equal to the reference value with very small error, the circuits connected with VSI-Gr, the load, and the grid are removed in the model **Hydro_PMSG_1a**, and the source 1200 V is set in the DC link instead. This replacement speeds up simulation roughly twice and makes investigation of an influence of the PMSG, turbine, and control system parameters easier.
Measuring water flow is not always possible with a sufficient degree of accuracy; therefore, the control methods that do not require knowledge of Q values are of interest. Such methods include search methods, for example, a perturbation and observation (P&O) method. At this, the measurement of the PMSG output power $P(k)$ under some value of the system input parameter, for instance, the rotational speed $\omega(k)$, is performed, and afterward it is

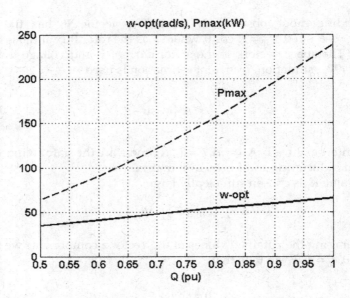

FIGURE 5.1
Dependences of the optimal rotational speed and the power on Q for the model **Hydro_PMSG_1**.

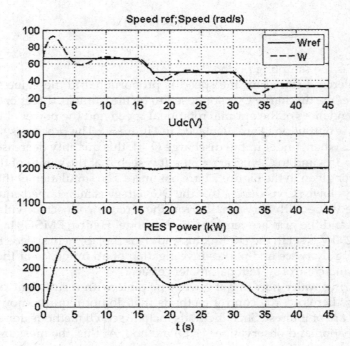

FIGURE 5.2
Processes in the model **Hydro_PMSG_1**.

decided, in which direction and how much the value ω has to be changed, in order to increase the value *P*. The algorithm may be written as

$$\omega(k+1) = \omega(k) + \Delta\omega(k)\,\text{sign}\big[\omega(k) - \omega(k-1)\big]\,\text{sign}\big[P(k) - P(k-1)\big]. \quad (5.6)$$

The value of the step $\Delta\omega(k)$ can depend on the value of the difference [*P*(*k*) − *P*(*k* − 1)] or can be constant. Such a method is utilized extensively in the photovoltaic systems, but in them it is a purely electrical one, whereas in the hydro turbines it is an electromechanical method demanding rotational speed change that leads to additional energy consumption; fortunately, the change of the water discharge takes place very slowly, so that the search process can be fulfilled rather rarely, for instance, with a period of some hours.

The realization of this algorithm is demonstrated in the model **Hydro_PMSG_2**, which is the revised model **Hydro_PMSG_1**, in which subsystem **Estimation** is replaced with the subsystem **Control**; this subsystem is taken from the model **Photo_m4S**, Ref. [3].

The subsystem **Control** is made with the employment of the blocks **Sample&Hold**. When input *S* = 1, the output follows input In; when *S* = 0, the output is held. The system operations are synchronized with the pulse generator; the pulse period is 20 s, because the steady state reaches rather slowly. With the pulse leading edge, the values *P*(*k*) and ω(*k*) and also sign [ω(*k*) − ω(*k* − 1)] and sign [*P*(*k*) − *P*(*k* − 1)] are stored. The value $\Delta\omega(k) \times$ sign [ω(*k*) − ω(*k* − 1)] × sign [*P*(*k*) − *P*(*k* − 1)] is formed. With the pulse trailing edge, the value ω(*k* + 1) is fabricated and stored, which will be utilized during the next sampling period. By this edge, with some delay, the values *P*(*k*) and ω(*k*) are rewritten in the blocks *P*(*k* − 1) and ω(*k* − 1) for utilization for the next calculation. The possibility to change (to increase) the iteration step $\Delta\omega(k)$ when the rate of power change is rather large is intended in the system. This opportunity is not used in the considered model; the step Δω is taken invariable and is equal to 5% of the nominal rotational speed.

Figure 5.3 shows the process when, at the initial value of *Q* = 1, this value decreases to *Q* = 0.75 at *t* = 140 s. The fluctuations in rotational speed are seen when the search system is running. Average values of speed and power are in keeping with the optimal values; changes in speed are accompanied by small short-term deviations of power. Note that to reduce the simulation time, the VSI is simulated in the *average-model* mode.

Instead of direct speed change, an indirect change of the rotational speed by alternation of the PMSG current component I_q can be utilized. At this, it is reasonable not to use the rotor position θ; instead, the voltage space vector **u** is employed with a shift by π/2. Instead of ω, the quantity I_q is used in relationship (5.6). This system is simulated in the model **Hydro_PMSG_3**. The same process as in Figure 5.3 is shown in Figure 5.4; the step of ΔI_q is 18 A, that is, about 5% of the nominal value.

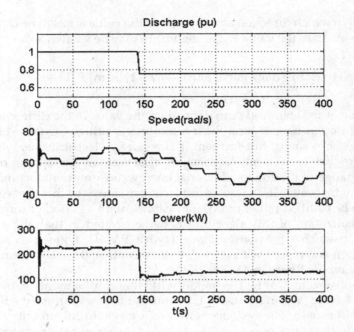

FIGURE 5.3
Processes in the model **Hydro_PMSG_2** with P&O method, ω change.

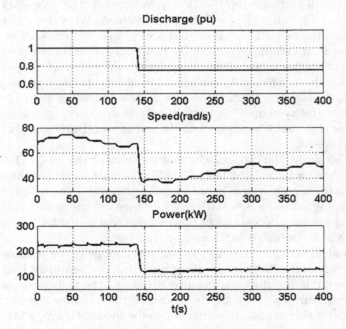

FIGURE 5.4
Processes in the model **Hydro_PMSG_3** with P&O method, I_q change.

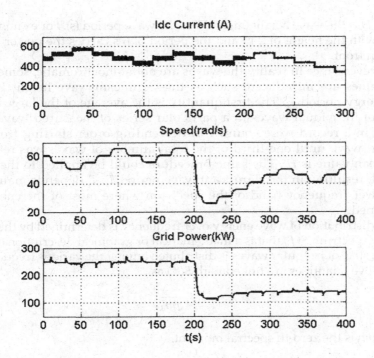

FIGURE 5.5
Processes in the model **Hydro_PMSG_4** with P&O method, ΔI_{dc} change.

In the system with an indirect change of the rotational speed, VSI-Ge may be replaced with the diode rectifier and the DC/DC boost converter; the input current ΔI_{dc} of the DC/DC converter is used for perturbation. The same structure is employed in the model **Wind_PMSG_9N** (Chapter 2). This system is simulated in the model **Hydro_PMSG_4**. The hydro turbine has a power of 250 kW under $Q = 1$ at the rotational speed of 60.3 rad/s. The process is shown in Figure 5.5, when Q changes from $Q = 1$ to $Q = 0.75$ at $t = 200$ s. The step of ΔI_{dc} is equal to 48 A, that is, about 10% of the nominal value. It is seen that, when $Q = 0.75$, the power reaches 140 kW, whereas at the nominal, invariable speed it would be only 125 kW.

5.2 Ocean Wave Energy Conversion

5.2.1 Point Absorber

It is assumed that the power of the monochromatic wave with the front length of 1 m is

$$P \approx 0.5 H^2 T (\text{kW/m}), \tag{5.7}$$

where H is the wave height (m) and T is the wave period (s). For example, the wave with the height of 2 m and the period of 6 s carries the power 12 kW per 1 m front.

However, since in reality the waves are not monochromatic, some average values are introduced, such as the significant wave height H_s and the energy period T_e. The first quantity is the average of the largest one-third of individual waves in a particular series of measured waves: the waves in a record were counted in descending order starting from the highest wave until one-third of the total number of waves was reached; the mean value is H_s. The same procedure could be applied to the wave period, resulting in the significant wave period T_e. T_e depends mainly on the lower frequency band of the spectrum where most of the energy is contained.

The distribution of wave energy over frequency is determined by the wave energy spectrum $S(f)$ that is a mathematical or graphical description of how a wave state of irregular waves is distributed among the various frequencies. The above quantities are linked with $S(f)$ as

$$H_s = 4\sqrt{m_0},\tag{5.8}$$

where m_0 is the zero-th spectral moment,

$$m_0 = \int_0^\infty S(f)\,df;\tag{5.9}$$

$$T_e = m_{-1}/m_0\tag{5.10}$$

$$m_{-1} = \int_0^\infty f^{-1}S(f)\,df.\tag{5.11}$$

Then

$$P \approx 0.5H_s^2 T_e\;(\text{kW/m}).\tag{5.12}$$

The value of T_e lies within 4–25 s, Refs. [5–7].

There are various expressions for the spectrum of sea waves. In particular, the rather popular spectrum of Bretschneider-Mitsuyasu is described by the ratio

$$S(f) = 0.257\frac{H_s^2}{T_e^4 f^5}\exp\left(\frac{-a}{(T_e f)^4}\right),\tag{5.13}$$

where the values 0.75–1.03 can be found for the quantity a in the references. Once the spectrum has been defined, the wave elevation of the i-th component wave can be found by

$$\eta_i = \sqrt{2S(f_i)\Delta f},$$ (5.14)

where Δf is the frequency spacing of the spectrum. Using each component wave elevation and frequency, the overall irregular wave form can be described as

$$\eta = \sum_i \eta_i \cos(k_i x - \sigma_i t - \varphi_i),$$ (5.15)

where φ_i is the random phase of the i-th wave component chosen from a uniform distribution on the interval $[0, 2\pi]$, $k_i = 2\pi/L_i$, where L_i is the space length of i-th wave component, $\sigma_i = 2\pi/T_i$, T_i is the period of this component. The values of L_i and T_i are linked with the relationship $2\pi L_i = gT_i^2$.

Another relationship gives the spectrum of Pierson–Moskowitz as

$$S(f) = \frac{0.31H_s^2 f_m^5}{f^5} \exp\left[-1.25\left(\frac{f_m}{f}\right)^4\right],$$ (5.16)

where f_m is the peak frequency (the spectrum has a maximum).

The energy of ocean waves is distributed unevenly in the world ocean. So, in the Atlantic Ocean, annual average wave power levels along the edge of North America's eastern continental shelf range from 10 kW/m to 20 kW/m. By comparison, shelf-edge wave power levels off the European western coastline increase from about 40 kW/m off Portugal up to 75 kW/m off Ireland and Scotland, decreasing to 30 kW/m off the northern part of the Norwegian coast. In the southern hemisphere, annual power levels of incident waves decrease with decreasing latitude, averaging more than 100 kW/m just south of New Zealand, and dropping to 30–40 kW/m in deep water west of Auckland, Ref. [6].

However, the use of existing huge reserves of energy is a very difficult task due to their variability. Wave power levels change at all timescales, from wave to wave (order of seconds), over periods of hours to days, from season to season and year to year. In addition, the wave-like nature of the energy itself creates certain difficulties in its extraction (as opposed to wind energy, which is almost constant in the steady state). In addition, issues of survivability under storm conditions are very important. This technology is far from a final decision; a wide variety of methods and devices for their implementation are being developed and researched by numerous groups of scientists and engineers. Practically there are no commercial installations. A wide variety of methods for extracting wave energy makes it difficult to create adequate models. The structure and parameters of the models are determined by the details of the structures and hydrodynamic processes that require knowledge and application of the theory of ocean waves, Ref. [8].

Due to the lack of established solutions and generic structures, the models considered further are approximate and are intended to familiarize with the problems that the engineer or the research electrician will encounter when creating the electrical part of such systems. As the most promising, currently, point absorbers (PAs) and oscillating water columns (OWCs) are considered.

The PA consists of a heaving buoy and a rod connected with the buoy; with the passage of a wave, the rod makes a reciprocating motion, which can be converted into electrical energy. In addition to movement in the vertical direction, the rod moves in the direction of the wave, which can also serve as a source of additional electricity, but these movements are not considered here.

Both the usual rotating generators and the linear generators can be employed in the PA. In the first case, it is necessary to have a converter of reciprocating motion into rotational one. A rack gear is used for this purpose, as shown in Figure 5.6, Ref. [9]. With the passage of the wave crest, the direction of rotation of the generator is reversed, which must be taken into account when designing the output electronics. By complicating the kinematics, it is possible to obtain an irreversible rotation.

Instead of mechanical transmission, the hydraulic one can be used (Figure 5.7). The hydraulic system contains four check valves that "straighten" fluid flow (hydraulic oil), so that the hydraulic motor at the outlet of the system, with which the electric generator is coupled, rotates in one direction.

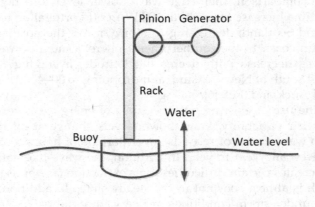

FIGURE 5.6
PA with a rack gear.

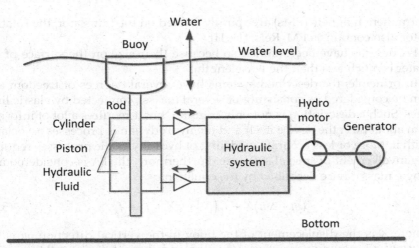

FIGURE 5.7
PA with a hydraulic transmission.

The direct coupling between the rod and the linear generator is shown in Figure 5.8. This technical solution seems to be the most natural, but the creation of a linear generator with good technical and economic characteristics is not an easy task. In the literature mentioned above, it is proved that at present such generators can be recommended for low power systems. It seems promising to use magnetic transmission, in which a large number of

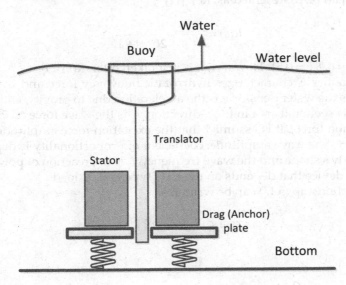

FIGURE 5.8
PA with a linear generator.

permanent magnets (PMs) are spirally located on the translator; the rotating rotor also contains a PM, Refs. [10, 11].

PA devices have been named so because their size on the surface of the water is much less than the wavelength.

In principle, the described systems have several degrees of freedom and can be considered as consisting of several masses connected by elastic linkages. Such a description is not only complex, but it requires a lot of information about both the device itself and the hydrodynamic processes associated with it. More or less accurate modeling of hydrodynamic processes requires the involvement of special software. Furthermore, the PA is considered as a single mass device described by the equation

$$(m + \Delta m)\ddot{X} + f_r\left(\dot{X}\right) + CX = f_w - f_g, \tag{5.17}$$

where X is the displacement of the buoy in the vertical direction, m is the mass of all moving parts, and Δm is the added mass of water that rises or falls with the buoy, this value depends on the frequency of the wave and is determined by hydrodynamic calculations, it is taken as the constant hereinafter: $M = m + \Delta m$; the term f_r is the radiation force, which is produced by an oscillating body creating waves on an otherwise calm sea, and for the Laplace transform $F_r(s)$ can be written as

$$F_r(s) = R(s)\dot{X}(s). \tag{5.18}$$

The transfer function $R(s)$ is the result of hydrodynamic calculations, in particular, it can be represented as, Ref. [12]

$$R(s) = \frac{B}{T_0^2 s^2 + 2\xi T_0 s + 1}, \tag{5.19}$$

where $T_0 = 0.9$, $\zeta = 0.51$. However, it is just taken often $R(s) = B$.

The quantity C characterizes hydrostatic buoyancy force and is $C = \rho g S$, where ρ is the water density, g is the acceleration due to gravity, and S is the buoy cross-sectional area in the x-direction. f_w is the wave force (also known as excitation force). It is assumed that the excitation force amplitude is proportional to the wave amplitude, coefficient of proportionality is dependent on the body's shape and the wave frequency, f_g is the reaction of power takeoff (PTO) device that depends on the PTO type and its load.

The relationship (5.17) can be written as

$$\ddot{X} = (f_w - f_g - CX - B\dot{X})/M = \frac{\Delta}{M}. \tag{5.20}$$

Then

$$\Delta = f_w - f_g - CX - B\dot{X} = f_w - f_g - \frac{C\Delta}{Ms^2} - \frac{B\Delta}{Ms}, \tag{5.21}$$

where s is Laplace transformation symbol. Suppose that the force f_g is proportional to the current of the driven generator; this value is $I = E/R$, where E is the generator electromotive force (EMF) and R is the load resistance.

Suppose, for definiteness, that a gear rack with wheel radius R_k is applied. When using the PMSG, we denote $K_e = K_i Z_p \Psi_r$, where $K_i = 1.5/R_k$, Z_p is the number of pole pairs, and Ψ_r is the rotor flux linkage. The phase voltage amplitude is $E = K_e \dot{X}/1.5$, the generator reaction force $f_g = K_e I = 0.666\, K_e^2 \dot{X}/R$, where R is the load phase resistance.

Therefore, we have the equations

$$\Delta = f_w - \frac{C\Delta}{Ms^2} - \frac{B\Delta}{Ms} - 0.666\frac{K_e^2 \Delta}{MsR} \qquad (5.22)$$

$$\Delta = \frac{f_w Ms^2}{Ms^2 + C + B_1 s} \qquad (5.23)$$

$B_1 = B + 0.666 K_e^2/R$

$$\dot{X} = \frac{f_w s}{Ms^2 + C + B_1 s} \qquad (5.24)$$

$$E = \frac{0.666 K_e f_w s}{Ms^2 + C + B_1 s} \qquad (5.25)$$

$P = 1.5E^2/2R$, where P is the average power.

For the harmonic signal with an angular frequency ω and amplitude f_{wa}

$$E = \frac{0.666 K_e \omega f_{wa}}{\sqrt{B_1^2 \omega^2 + (C - M\omega^2)^2}}. \qquad (5.26)$$

The program **PAm1.m** calculates the dependence of the generator power on the load resistance according to the above formulas. Figure 5.9 shows the results for the parameters PA and the generator used in the model PA1.mdl: $M = 0.42 \times 10^5$, $C = 0.19 \times 10^6$, $B = 0.91 \times 10^4$, $f_{wa} = 0.3 \times 10^6$, $K_i = 12$, $Z_p = 18$, and $\Psi_r = 1.429$. It can be seen that with a certain load resistance, the maximum of the power produced by the generator is reached (see, e.g., Refs. [13, 14]).

Program PAm1.m

```
F=3e5;
w=0.8
C=0.19e6;
m=0.42e5;
b=0.91e4;
M=C/m;
Ke=1.5*18*1.429/0.125;
```

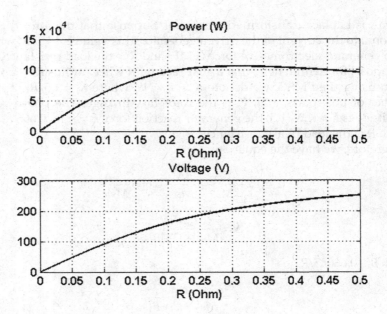

FIGURE 5.9
Dependence of the PA power from the load resistance.

```
for i=1:500;
R(i)=0.001*i;
L(i)=(b+0.666*Ke*Ke/R(i))/m;
E1(i)=(w^2-M)^2+L(i)^2*w^2;
E(i)=0.666*Ke/sqrt(E1(i))*F*w/m
P(i)=1.5*E(i)^2/R(i)/2
end;
subplot(2,1,1)
plot(R,P)
grid
xlabel('R (Ohm)')
title('Power (W)')
subplot(2,1,2)
plot(R,E)
grid
xlabel('R (Ohm)')
title('Voltage (V)')
```

The maximum of the average power is 105 kW and is obtained with $R = 0.31\ \Omega$ when the phase voltage amplitude is 210 V. The considered system is simulated in the model **PA1** with the PMSG. It is easy to verify, by repeating the simulation with a change in load resistance, that with a phase resistance of 0.31 Ω, the maximum average mechanical power of 0.105 MW will be reached. The generator output voltage curve is shown in Figure 5.10. When the disturbance frequency changes, the value of the optimal resistance also

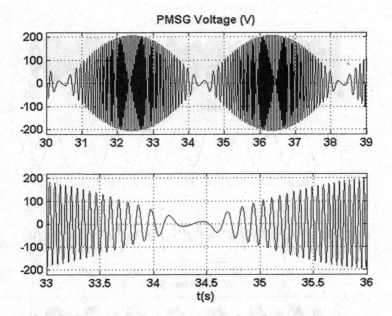

FIGURE 5.10
PA generator voltage in the model **PA1**.

changes. For example, when the frequency decreases from $\omega = 0.8$ rad/s to $\omega = 0.6$ rad/s, the optimal load impedance decreases to 0.22 Ω with the maximum average power of 75 kW.

The model **PA2** simulates the same system as the model **PA1** but with DC output. It was found that the optimal value of the load resistance is 0.55 Ω; the average generator power is 108 kW.

However, usually the PA output is the AC voltage transmitted to the grid. In this case, it is not the load resistance that is tuned, but the generator current corresponding to this load resistance at the actual generator voltage. For this purpose, the DC/DC boost converter and VSI can be used, Ref. [13]. Such a system is investigated in the model **PA3c**. The current of the reactor of the DC/DC boost converter is defined as the ratio of the EMF of the generator to the value of the optimal resistance. The control circuits of the VSI-Gr are usual; the specified voltage value of the capacitor at the input of the inverter is 1000 V.

If to perform a simulation, one can see that the average value of the power delivered to the grid is 100 kW, while its instantaneous values fluctuate within 0–200 kW.

Most probably, such wave energy converters (WECs) will not be used alone, but as WEC farms; then, with parallel and (or) series connection of a number of WECs arranged arbitrarily relative to the wave crest, the resulting power will be smoothed. In the model **PA3**, three WECs are applied, which

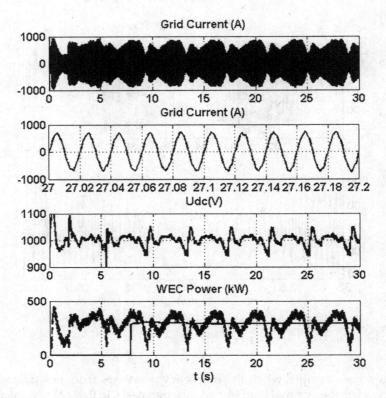

FIGURE 5.11
Processes in the model **PA3**.

are evenly distributed relative to the direction of the wave and connected in parallel. The DC-to-DC and VSI converters are common to all WECs. Figure 5.11 shows the grid current, instantaneous and average value of the power delivered to the network, as well as the capacitor voltage. The average power is 292 kW, the fluctuations of the instantaneous values are reduced significantly.

Model **PA3b** is the same as **PA3**, but the wave is non-monochromatic, the amplitude ratios of its individual components correspond to the data given in Ref. [15] (Table 5.1). To obtain interesting results, the simulation time should be significantly increased. Figure 5.12 shows the power delivered to the grid over a period of 120 s, as well as the average value of this quantity with an averaging period of 7.85 s.

Utilization of the active rectifier instead of DC/DC converter gives more opportunities for system optimization. Three WECs are connected in parallel in the model **PA4b**, as in the model **PA3**, the DC link voltage is kept at the designated level of 1000 V. The control system implements the regulation of the generator current component I_q, which determines the power generated

TABLE 5.1

Non-Monochromatic Wave Parameters

Component	Amplitude (m)	Frequency (Hz)	Phase (rad)
a0	0.455	0.266	−2.2
a1	0.915	0.198	−0.942
a2	1.085	0.151	0.628
a3	2.09	0.115	−π
a4	1.19	0.077	−2.04

by the generator, the reference for the component I_q is determined in the same way as in the previous model. The schematic and parameters of each WEC are also the same as in the previous model.

It is assumed that VSI-rectifiers are located near the generators, and VSI-inverter and devices for linkage with the grid are located on the coast at a distance of 10 km and are connected to WECs by a submarine cable. The voltage in the DC link is assumed to be 1 kV with the inverter output voltage of 380 V and the grid voltage of 6.3 kV, for which a transformer with the appropriate voltage levels is used.

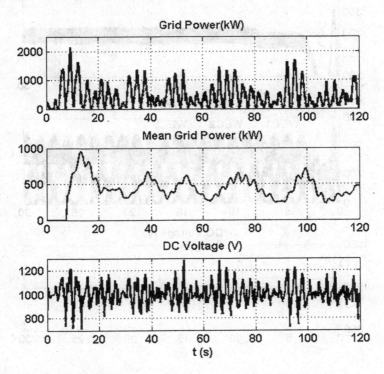

FIGURE 5.12

Processes in the model **PA3b** with non-monochromatic wave.

Note that at the outputs of the generators, the resistors with a relatively small resistance of 10 Ω are placed, and when trying to increase the resistance values, the simulation process slows down dramatically. These resistors distract appreciable power. Since in reality these resistors are absent, the power consumed by them is measured (*LossP*, only in WES1), and then added to the output power of the system in order to get the correct picture of the power flow.

Figure 5.13 shows the curves of the grid and individual powers of WECs, and DC voltage in the line as well. The average active power is 300 kW. The shape of the grid current is close to a sinusoid, and the capacitor voltage is kept with a sufficient degree of accuracy. Power pulsations are less than that in the previous model, and they will decrease with the increasing number of parallel-connected generators.

Existence of the output power ripples stimulates the use of additional tools to smooth it out. For example, it is promising to use a supercapacitor for this

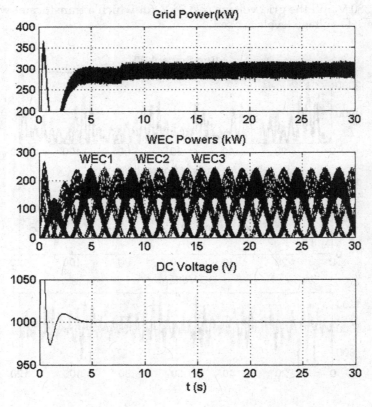

FIGURE 5.13
Processes in the model **PA4b**.

purpose. The model **PA32** is a modification of the model **PA3**, in which the subsystem **UC** is added that contains a supercapacitor, which, with the help of contactless switches, connects to DC link voltage U_{dc}. When $U_{dc} > 1025$ V, the capacitor is connected in the charge mode, absorbing excess power. When U_{dc} reduces to 1000 V, the capacitor is turned off. With further reduction of U_{dc} to 975 V, the capacitor is connected in the discharge mode, supplying additional power. The banks of supercapacitors with the parameters 5.8 F, 160 V, and series resistance of 0.24 Ω are used, Ref. [16]. Seven banks are connected in series in 11 parallel branches, so that the total capacity is 5.8 11/7 = 9.1 F. Figure 5.14 shows the changes in grid power and DC voltage. From comparison with Figure 5.11, it is clear that the fluctuations of the mentioned quantities have noticeably decreased.

Along (and more often) with a rigid connection of the rod with the generator, other methods are used. As already mentioned, Refs. [10, 11] consider the use of magnetic transmission. The device consists of two main parts: a rotor and a translator. The translator is to be coupled to the buoy to move up and down along its axis. The rotor, to be coupled to a rotary machine, rotates about the same axis. Each part is made of a ferromagnetic core that supports

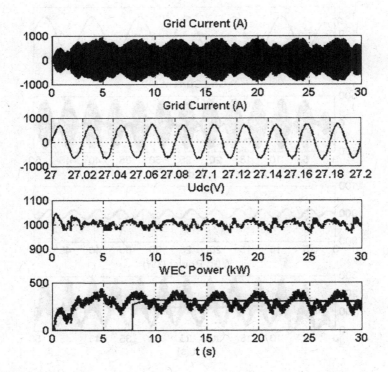

FIGURE 5.14
Processes in the model **PA32** with supercapacitor.

two alternating helically shaped PM poles. The gear ratio G is defined as $2\pi/\lambda$ in which λ is the width of helix. The transfer torque is

$$T_r = T_m\sin(\theta - GX),\tag{5.27}$$

where θ is the generator rotation angle, X is the rod travel, T_m is the maximum of the transfer torque that is equal to F_m/G, and F_m is the stall force. Such a system is simulated in the model **PA5**. The parameters of the transmission are taken from Ref. [11]: $F_m = 500$ кH with $\lambda = 44$ mm, so that $G = 2\pi/0.044 = 142.7$. The PA and PMSG parameters are as follows: $M = 0.3 \times 10^6$, $C = 0.56 \times 10^6$, $B = 0.91 \times 10^5$, $f_{wa} = 0.5 \times 10^6$, $Z_p = 4$, and $\Psi_r = 1$ Wb. Unlike PA previous models, in which the generator was simulated in the mode *"Mechanical input: Speed W,"* in the model under consideration, it is simulated in the mode *"Mechanical input: Torque Tm."* With the parameters taken, the optimum phase load resistance is approximately 1 Ω. The voltage in the DC link is assumed to be 1000 V. The generated electricity is transmitted to the 380 V grid using VSI-Gr. The average power delivered to the grid is 101 kW. The processes in the system are shown in Figure 5.15.

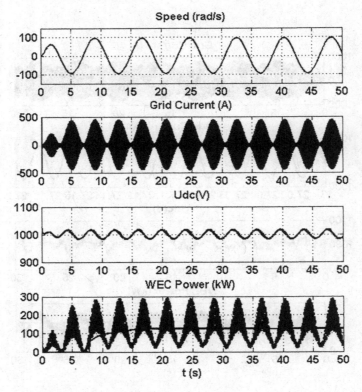

FIGURE 5.15
Processes in the model **PA5** with magnetic transmission.

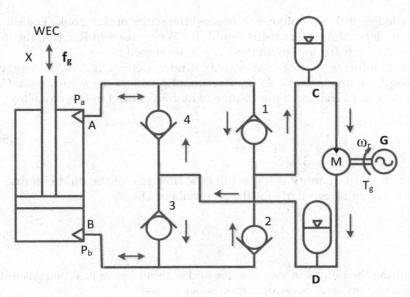

FIGURE 5.16
Schematic of the hydraulic system for conversion of reciprocating motion into the rotary one.

Hydraulic systems are often used to convert the reciprocating motion of the rod into the generator rotary motion. An additional advantage of the hydraulic system is its ability to store energy for a short time and then use it, which smooth out torque fluctuations caused by the wave nature of the disturbing force. The structure of the commonly used hydraulic system is depicted in Figure 5.16.

The WEC rod moves the piston of an oil-filled hydraulic cylinder that has two outlets near the ends of the cylinder. With the help of four unidirectional valves, the flow direction of the fluid is straightened. Let, for example, the piston moves up. Then the fluid pressure at point A increases, under the action of this pressure, the fluid enters the gas-filled high-pressure accumulator C through valve 1. Some of this fluid is sent to the hydraulic motor M, and some remains in the accumulator, while the gas contained in it is compressed. When the pressure of the incoming fluid weakens, the gas expands and directs the fluid stored in the accumulator into the hydraulic motor, maintaining the pressure in it, reducing the torque fluctuations of the hydraulic motor. After the hydraulic motor, the fluid, through the low-pressure accumulator D and the valve 3, returns to the hydraulic cylinder. When the rod moves down, the process proceeds in a similar way, but the fluid passes through the valves 2 and 4.

The models that take into account, inter alia, the detailed characteristics of unidirectional valves are developed in Refs. [12, 17]. The toolboxes MATLAB® *SimHydraulics* and *SimMechanics* are used for model building in Ref. [18];

knowledge of these toolboxes is beyond the scope of this book. The simplified model of the system using Simulink® is proposed in Ref. [19], the main ideas of which are applied in the models developed here.

The volumetric flow of fluid from a cylinder is equal to $\dot{V}_p = A_p v$, where A_p is the piston area and $v = dX/dt$. This quantity enters the accumulator C, so that the rate of change of the volume of hydraulic fluid in C is equal to

$$\frac{dV_c}{dt} = \dot{V}_p - \dot{V}_m , \tag{5.28}$$

where the last quantity is the volumetric flow rate of the fluid entering the hydraulic motor. Similarly for the accumulator D

$$\frac{dV_d}{dt} = \dot{V}_m - \dot{V}_p , \tag{5.29}$$

assuming the fluid is incompressible and without losing it. When gas (nitrogen) is compressed, polytropic ratio is performed:

$$P_N(0) V_N(0)^{1.4} = P_N(t) V_N(t)^{1.4} , \tag{5.30}$$

the initial values of the gas pressure and volume are in the left side. The rate of change in the volume of gas must be equal and opposite in sign to the rate of change in the volume of hydraulic fluid:

$$\frac{dV_{Nc}}{dt} = -\frac{dV_c}{dt} , \tag{5.31}$$

$$\frac{dV_{Nd}}{dt} = -\frac{dV_d}{dt} . \tag{5.32}$$

Thus, it is possible to find the difference of accumulator pressures ΔP. Then the reaction force of the PTO is

$$f_g = \Delta P A_p sign(v). \tag{5.33}$$

The hydraulic motor with a constant rotation speed ω_r and a variable displacement V_m is considered in Ref. [19]. The torque of the hydraulic motor T_m is determined by the torque of the attached generator T_g, which is assumed to be known. Then the required flow rate

$$\dot{V}_m = \frac{dV_m}{dt} = \frac{T_g \omega_r}{\Delta P} . \tag{5.34}$$

Moreover, this value used in the above ratios has limitations on the minimum and maximum. In other cited works, a variable-speed hydraulic motor with a constant volume D_m is considered. Hydraulic motor torque is

$$T_m = D_m \Delta P / 2\pi \tag{5.35}$$

and may differ from the generator torque since the relationship

$$J\frac{d\omega_r}{dt} = T_m - T_g \tag{5.36}$$

is carried out in the transients, where J is the total moment of inertia of the hydraulic motor and attached generator. The value of T_g is often taken to be proportional to the speed of rotation:

$$T_g = k_g \omega_r. \tag{5.37}$$

Value

$$\dot{V}_m = \frac{dV_m}{dt} = D_m \omega_r / 2\pi. \tag{5.38}$$

Thus, we arrive at a model whose scheme is shown in Figure 5.17; the quantity $\dot{V}_m$ shown in Figure 5.17 is equal either to Equation (5.34) with $\Delta P = P_c - P_d$ or to Equations (5.35), (5.36), and (5.38). This scheme is implemented in the model **PA6s**. Model parameters are borrowed mainly from Ref. [17]. Here $A_p = 0.0235$ m^2 and $D_m = 500$ sm^3; for the high pressure accumulator, $P_N(0) = 13.2$ MPa with the volume of $V_N(0) = 7.5$ m^3; and accordingly 6.6 MPa and 3 m^3 for a low pressure accumulator.

It is shown further how to find steady-state, more precisely, quasi-steady-state values, assuming that the rate of change of the input signal is a pure sinusoid with amplitude A_w as in the model under consideration.

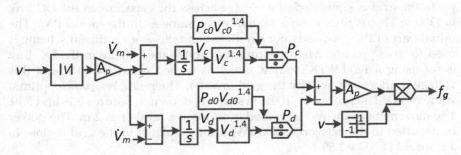

FIGURE 5.17
Block diagram of the hydraulic PTO system.

The constant component of the quantity dV_p/dt is $2/\pi\, A_w\, A_p$ and must be equal to $D_m\omega_r/2\pi$ from which the value ω_r can be found. In the case, $A_w = 1$ and

$$\omega_r = \frac{2 \times 1 \times 2.35 \times 10^{-2} \times 2 \times \pi}{\pi \times 500 \times 10^{-6}} = 188 \text{ rad/s.}$$

Let $T_m = T_g$ and $k_g = 24$. Then
$T_m = D_m\Delta P/2\pi = T_g = 24 \times 188$, from which

$$\Delta P = \frac{24 \times 188 \times 2 \times \pi}{500 \times 10^{-6}} = 56.67 \text{ MPa.}$$

It has to be

$$\frac{13.2 \times 7.5^{1.4} - 6.6 \times 3^{1.4}}{V_0^{1.4}} = 56.67,$$

therefore $V_0 = 2.34$ m³. This value has to appear at the outputs of the integrators. The PTO reaction is $f_g = 56.67 \times 0.0235 = 1.33$ MN.

We find the values of resistance R at the output of the generator to obtain the value of $k_g = 24$. Amplitude of the generator phase voltage is $E_a = Z_p\Psi_r\,\omega_r = 4 \times 1 \times \omega_r$. Then the three-phase power is $P = 1.5E_a^2/R$, or $T_m = 1.5 \times 4^2\,\omega_r/R$, therefore $R = 1.5 \times 16/24 = 1\ \Omega$.

The given parameters are used in the model **PA6s**. After completing the simulation, you can verify that the model of the hydraulic system functions as required.

In the model **PA6**, the considered hydraulic system is driven by the WEC, the parameters of which are also mainly taken from Ref. [17]. At $t = 300$ s, the wave amplitude increases from 1 m to 1.5 m. The generator is loaded on a 3-Ω phase resistance. Its power increases from 200 kW to 460 kW, while power fluctuations are insignificant. The reaction force of the PTO increases from 380 kN to 590 kN.

In the model **PA6b**, the transfer of electric power generated by the generator to the grid is simulated, the voltage across the capacitor in the DC link is 1200 V. The rectifier control system is the same as in the model **PA5**. The initial part of the process is not simulated; therefore, an artificial scheme is used for this process: when starting the model, the voltage in the DC link is set using a 1200-V DC source, which is turned off at $t = 30$ s, and this voltage is afterward kept by the grid inverter. The phase resistance optimal value is 3 Ω. The processes in the system are shown in Figures 5.18 and 5.19. The movement of the rod is limited by stops at a level of ±1.2 m. The power transmitted to the network is 200 kW, and the current in the grid is close to a sinusoid (THD $\approx$ 2.5%).

Since the rod makes a reciprocating motion, it seems reasonable to employ a linear generator, which does not require utilization of the intermediate

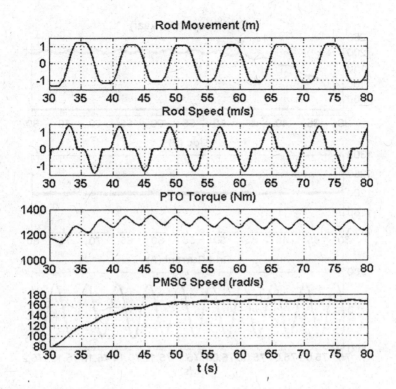

FIGURE 5.18
Processes in the PA of the model **PA6b**.

devices and increases the reliability of the installation, Refs. [20–23]. The disadvantages of a linear generator include a sufficiently large value of the leakage inductance of the windings and poor weight-dimensional features: for example, a 10-kW small power generator has a mass of 1 ton, Ref. [22].

In a conventional design, the stator windings are stationary, and the PMs are located on the translator moving in reciprocating motion. If to denote as τ_p, a pitch of the location of the magnets with alternating polarity (pole pitch), then the induced voltage can be represented as

$$E = \frac{\pi V}{\tau_p} \Psi_r \cos\left(\frac{\pi X}{\tau_p}\right),$$ (5.39)

where V is the translator movement speed.

It can be written for the model **PA1**, in which the generator is connected directly to the rod

$$E = K_i \Psi_r Z_p V \cos(K_i Z_p X).$$ (5.40)

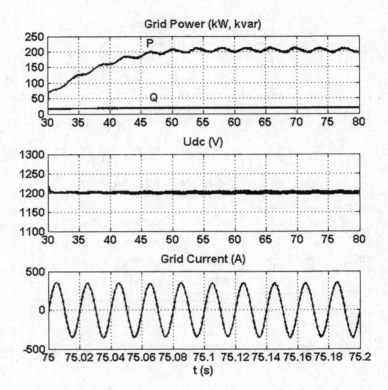

FIGURE 5.19
Processes in the WEC of the model **PA6b**.

Since the equations describing conventional and linear generators with surface-mounted PMs are the same, then the **PA1** model with the adequate parameters can be considered as a WEC with a virtual linear generator with $K_i Z_p = \pi/\tau_p$. Such a system is used in the model **PA7**. The generator parameters are taken mainly from Ref. [24]. It is accepted $\tau_p = 0.05$ m, $Z_p = 20$, and $K_i = 3.14$. The maximum average power is achieved with a load resistance of ~13 Ω/phase and is equal to 102 kW. This system can be applied in the previously discussed structures. For example, in model **PA7b**, the generator is controlled by the VSI, which regulates the current component I_q of the generator, as in model **PA4b**. The voltage in the DC link is assumed to be 3000 V. If to repeat simulation with different ratios between the speed of movement of the rod and the magnitude of the current demand, then it can be found that the maximum average power delivered to the grid is 75 kW and is achieved when the value of the said ratio is 25.6 × 3.14 As/m.

It is also possible to investigate the introduction of additional current reference signals depending on the acceleration or movement of the rod. If, for example, to add the current reference signal that is proportional to the acceleration of the rod, by setting the parameter *Mode* = 1 in the subsystem

Control, then, if the proportionality factor is set to 20, the average power delivered to the network will increase to 150 kW, but the maximum current value of the generator will also increase from 80 A to 200 A.

5.2.2 Oscillating Water Columns

The OWC is a chamber having underwater and surface parts: in the first part there is an opening for entry of water, and at the top of the topside an air turbine is installed. When the water level rises, the pressure on the air inside the chamber grows and becomes above the atmospheric pressure, the air passes through the turbine to the outside, which in this case is driven into rotation. After passing through the crest of the wave, the pressure inside the chamber begins to fall and becomes less than the atmospheric pressure; the atmosphere air flows through the turbine into the chamber, causing the turbine to rotate. The air turbine is used, whose direction of rotation does not depend on the direction of airflow (Wells turbine or impulse turbines). A generator is connected to the turbine shaft (Figure 5.20). To prevent damage during strong rough sea, an exhaust valve is placed in the chamber. An OWC may be placed in a fixed structure or a floating structure.

In the OWC, the conversion of wave energy into electrical energy is carried out in two steps: the conversion of wave energy into pneumatic energy in the OWC chamber and the conversion of the latter into mechanical energy, and then into electrical energy in the air turbine with a generator. In the literature, one can find two approaches as for investigation of the OWC. Evaluation and modeling of the first conversion stage requires detailed knowledge of the camera design; therefore, in a number of works only the second stage is investigated; the input value is the pressure drop in the

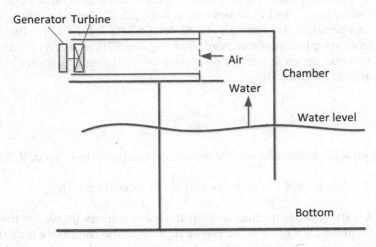

FIGURE 5.20
OWC construction.

turbine dP, using experimental or postulated values of this quantity. Such a method was used in Ref. [25]. In other studies, approximate relationships are used to estimate the airflow velocity V_x based on information about wave parameters, Ref. [26].

The following equations are valid for the Wells turbine, Refs. [25, 26]:

$$dP = C_a K(1/a)[V_x^2 + (r\omega_t)^2] \qquad (5.41)$$

$$T_t = C_t Kr[V_x^2 + (r\omega_t)^2] \qquad (5.42)$$

$$\phi = \frac{V_x}{r\omega_t}. \qquad (5.43)$$

Here, $K = \rho bln/2$, dP is the pressure drop across rotor (Pa), C_a is the power coefficient, a is the cross-sectional area (m^2), V_x is the airflow velocity (m/s), ω_t is the turbine angular velocity (rad/s), T_t is the torque produced by turbine (Nm), C_t is the torque coefficient, ϕ is the flow coefficient, b is the blade height (m), l is the blade chord length (m), n is the number of blades, r is the mean radius, and ρ is the air density.

Equation of motion of the rotor with the attached generator

$$J\frac{d\omega_g}{dt} = T_t/i - T_g, \qquad (5.44)$$

$\omega_t = \omega_g/i$, where i is the gear ratio and T_g is the generator torque, which is a known function of ω_g, this function is selected during system design; J is the moment of inertia of the rotating parts reduced to the generator shaft.

The coefficients C_a and C_t depend on ϕ, Refs. [25, 26]. In the working range, $C_a = f\phi$ can be taken, for example, $f = 8$, Ref. [26]. As for the C_t coefficient, its dependence on ϕ is essentially nonlinear (Figure 5.21), and at a certain value of ϕ the torque drops to zero (stop mode). Substituting Equation (5.41) into Equation (5.42), we obtain

$$T_t = \frac{C_t}{C_a} ardP. \qquad (5.45)$$

For option with V_x calculation, the following relationship is used, Ref. [26]:

$$V_x = \left(8Acw_c/\pi D^2\right) \times \sin\left(\pi l_c/cT\right) \times \cos\left(2\pi t/T\right), \qquad (5.46)$$

where A is the wave amplitude (m); c is the wave velocity (m/s); T is the wave period (s), at this, $T = \lambda/c$, λ is the wavelength; w_c is the chamber's inner width (m); D is the diameter of the turbine duct (m); and l_c is the capture chamber's length (m).

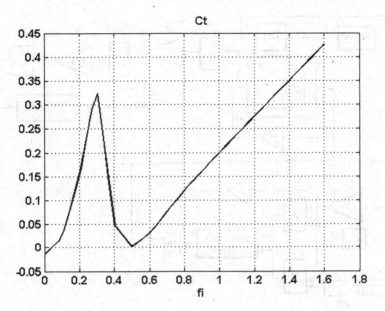

FIGURE 5.21
Dependence of the torque coefficient on the flow coefficient.

Because usually $\pi l_c \ll cT$, after replacement of sin with its argument, denoting $S_{ch} = w_c l_c$ that is the chamber cross-sectional area (m²), $S_d = \pi D^2/4$ is the duct cross-sectional area (m²), $S_o = S_{ch}/S_d$, $\omega = 2\pi/T$, we obtain

$$V_x = \omega A S_o \cos(\omega t). \tag{5.47}$$

Thus, to calculate V_x by this method, it is necessary to know some parameters of the OWC construction.

For the option with dP as an input quantity, the basic properties of the system in question can be quickly investigated without utilization of *SimPowerSystems* blocks, using the simulation scheme shown in Figure 5.22. The turbine torque is calculated by Equation (5.45), with C_t being dependent on φ using the table. The value of φ is calculated by Equation (5.43) at the known speed of rotation, obtained by solving Equation (5.44) with a known dependence of the generator torque on its rotational speed. The value of V_x is calculated by solving Equation (5.41), which, for known values of dP and ω_t, reduces to a third-order algebraic equation

$$V_x^3 + pV_x + q = 0,$$

$$p = r^2 \omega_t^2,$$

$$q = -\frac{ra\omega_t dP}{fK},$$

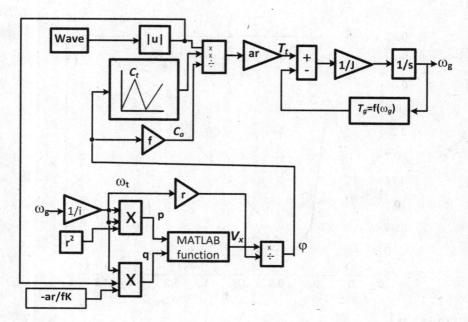

FIGURE 5.22
OWC block diagram.

whose solution by Cardano's formula is

$$V_x = \sqrt[3]{-\frac{q}{2} + \sqrt{\frac{q^2}{4} + \frac{p^3}{27}}} + \sqrt[3]{-\frac{q}{2} - \sqrt{\frac{q^2}{4} + \frac{p^3}{27}}}. \tag{5.48}$$

The power available from the airflow in the OWC chamber is, Ref. [25],

$$P_i = aV_x \left(dP + 0.5\rho V_x^2 \right). \tag{5.49}$$

The described system is simulated in the model **OWC1**. The model parameters are taken from Ref. [25]: $n = 8$, $K = 0.7079$, $r = 0.7285$, and $a = 1.1763$; it is accepted $f = 8$, $J = 10$ kg m². As a standard value, $dP = |\, 7000 \sin (0.1\pi t)\, |$ Pa is accepted. The linear or quadratic dependence of the generator torque on the rotational speed can be implemented.

Reiterating simulation, it can be verified that, with a linear relationship, the maximum average power of 33.5 kW is reached at $T_g = 3.2\omega_g$ when $\varphi = 0.3$. With a quadratic dependence, the maximum average power, equal to 38.4 kW, is reached at $T_g = 0.05\omega_g^2$; at this $\varphi = 0.3$ as well. Thus, it is really optimal to control the power of the generator in proportion to the cube of its rotational speed.

Another method to solve the cubic equation is used in the model **OWC1a**, namely, with the help of the closed loop system:

$$\hat{\varphi} = \int K_p \left(\frac{adP}{r^2 \omega_t^2 fK} - \varphi^3 - \varphi \right) dt, \tag{5.50}$$

where the bracketed expression is another notation for Equation (5.41).

Let's compute the value dP by Equation (5.41) with the use of Equation (5.47) for the OWC with parameters given in Ref. [26]: $K = 0.1$ kg/m, $S_d = 0.44$ m², $S_{ch} = 4.5 \times 4.3 = 19.35$ m², $r = 0.375$ m, $S_o = 44$, $A = 1.2$ m, and $T = 10$ s at $\varphi = 0.3$. We receive $V_x = 2\pi/10 \times 1.2 \times 44 \times \cos(\omega t) = 33.1\cos(\omega t)$ m/s, $r\omega_t = 33.1/0.3 = 110.3$ m/s, and $|dP| = (8 \times 0.3 \times 0.1/0.44) \times (33.1^2 + 110.3^2)$ that gives $dP = 7234 \cos(\omega t)$ Pa; it is consistent with the values assumed when using the dP reference option. Thus, to study the main characteristics of OWCs, it does not matter which of the two mentioned simulation methods is applied; furthermore, the method with the dP assigning will be used, since it is more universal, because it does not require knowledge of the parameters of the OWC chamber.

One can prove that with irregular waves, the system does not function satisfactorily, since with possible sudden change in pressure, the rotational speed can also change dramatically, the coefficient φ turns out outside the optimal zone, and with subsequent pressure recovery, the normal mode does not return. Therefore, it is necessary to limit possible changes in the speed of rotation of the turbine either by connecting the generator to the grid or by using a speed controller.

In the model **OWC1b**, the squirrel cage IG is connected to the grid with the voltage of 380 V. The 75-kW generator has two pairs of poles and is connected to the turbine via a step-up gearbox $i = 1.5$. The similar system was investigated in Ref. [27]. If to repeat the simulation with different values of A_p of the amplitude dP for the regular waves, it can be seen that the power delivered to the grid increases with A_p and reaches the maximum of 64.8 kW at $A_p = 10.5$ kPa and decreases with further A_p increase. Thus, it is desirable to limit the pressure on the turbine to this value, which can be achieved by installing an air valve; the model of such a valve depends on its construction. It is assumed in the model **OWC1b** that when the pressure exceeds 10.5 kPa, the pressure value decreases with a speed proportional to the excess pressure. The unit with the first-order transfer function takes into account the dynamics of the servo system and pressure meter. Decreasing the pressure below 10.5 kPa restores the pressure value corresponding to the input value.

It is accepted when a process with regular waves is simulated that at $t = 40$ s, the input pressure increases from 7 kPa to 14 kPa. The process without air valve is depicted in Figure 5.23. It is seen that the value φ can reach 0.4, and the turbine torque decreases essentially at these instants. The average power delivered in the grid is about 47.3 kW.

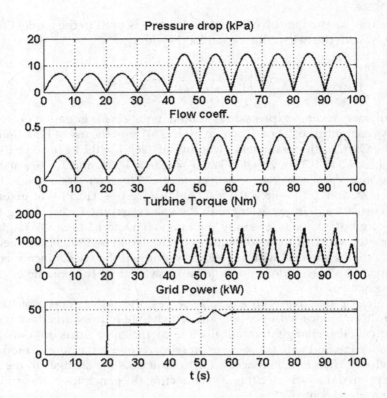

FIGURE 5.23
Processes in the model **OWC1b** under regular waves without air valve.

The process with the same pressure but with the air valve is shown in Figure 5.24. It is seen that the pressure drop in the turbine is not more than 11.5 kPa; the value of φ is limited by 0.3, the torque decrease is much less; as a result, the power sent to the grid reaches 73 kW.

The process with irregular waves is shown in Figure 5.25. Like in the previous model, the input pressure doubles at $t = 40$ s. Without the air valve, the power changes within 30–50 kW after this time. The same process but with the active valve is shown in Figure 5.26. The grid power rises noticeably during 40–100 s.

The possibility to control the rotational speed of the turbine appears with the use of doubly fed induction generator (DFIG), the rotor circuits of which includes two VSIs, as in the model **Wind_DFIG_1N**: VSI-Ge and VSI-Gr. Such a system is investigated in the model **OWC2**. The DFIG with a power of 75 kVA, 1000 rpm is used. The generator is connected to the turbine without a gearbox. The generator model in pu is utilized, while the WEC model operates in SI. The inverter control systems are borrowed from the above-mentioned

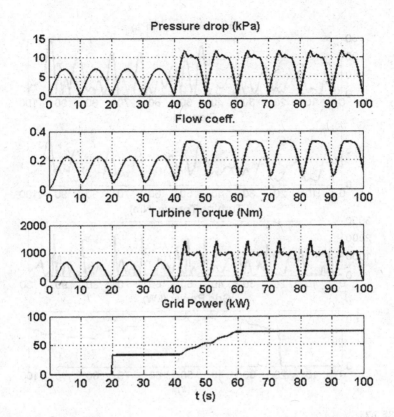

FIGURE 5.24
Processes in the model **OWC1b** under regular waves with air valve.

model; to reduce the simulation time, the direct torque control (DTC) system utilized to control VSI-Ge uses the rotor flux linkage values from the DFIG model, whereas their estimates are used actually. To reduce the duration of the simulation when it is reiterated (see below), the inverter circuits connected to the grid are removed in the model **OWC2a**, the generator power is found as a product of its electromagnetic torque and speed.

Carrying out simulation repeatedly with different dP amplitudes and rotational speeds ω, one can find the rotational speed ω_{opt} when the generator power is maximum P_{max} at the given value dP. If to connect these points, we get the curve shown in Figure 5.27, which, with a reasonable degree of accuracy, can be approximated with a third-degree parabola $P_{max} = K_m \omega^3$, where in this case $K_m = 0.813$. The processes in the model OWC2 under $A_p = 7$ kPa are shown in Figure 5.28.

Thus, it is desirable to adjust the speed of rotation of the turbine to obtain maximum power at the existing wave parameters. There are several ways to

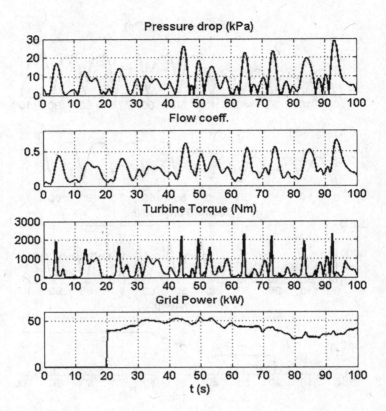

FIGURE 5.25
Processes in the model **OWC1b** under irregular waves without air valve.

implement optimal control. In Ref. [28], it is proposed to adjust the generator torque depending on the actual rotational speed as

$$T_g = K_m \omega^2. \tag{5.51}$$

This structure is implemented in the model **OWC2c** using the DTC system. In this case, however, it turns out that, with a significant pressure increase, the coefficient φ increases sharply, the torque of the turbine and its rotational speed drop sharply and the normal mode is not restored. Such a case is shown in Figure 5.29 with an increase in pressure amplitude from 7 kPa to 13 kPa at $t = 30$ s (Threshold of **Switch1** in the subsystem **Gen_Contr** is 0.5×10).

To eliminate this phenomenon, it is necessary to measure the pressure drop in the chamber. Then it is possible to calculate the coefficient φ using the above formula, and in case of exceeding the value of 0.3, set the zero torque of the generator (which is braking in this case) and thus prevent

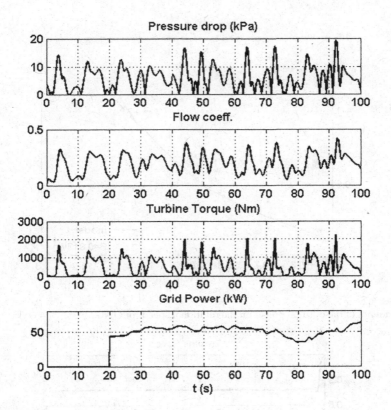

FIGURE 5.26
Processes in the model **OWC1b** under irregular waves with air valve.

braking of the turbine with the generator. It can be seen from the plot in Figure 5.30 (Threshold of **Switch1** in the subsystem **Gen_Contr** is 0.5) that the OWC operates normally, and the values of the generator power agree with the cubic dependence: if to consider the times when the speed and power reach maximum values, we have for $\omega = 1.05$, $P = 0.95$; for $\omega = 1.43$, $P = 2.4$ (it should be $P = 0.813 \times 1.05^3 = 0.94$ and $P = 0.813 \times 1.43^3 = 2.38$, respectively). Similar results can be obtained if to consider not instantaneous, but average values of speed and power. Figure 5.31 shows processes under irregular waves, the OWC operates normally; at the same time, it can be verified that without the proposed limitation, normal operation is not ensured.

Another approach is implemented in the model **OWC2b**. The system has a speed controller, whose reference is

$$\omega_{ref} = \sqrt[3]{\frac{P_{gen\_mean}}{K_m} + K_d \frac{dP_{gen\_mean}}{dt}}. \qquad (5.52)$$

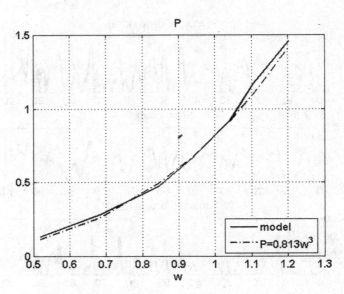

FIGURE 5.27
Dependence of the optimal power on the rotational speed for OWC.

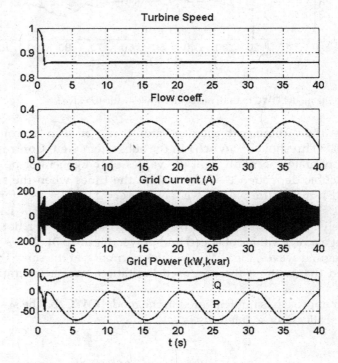

FIGURE 5.28
Processes in the model **OWC2**.

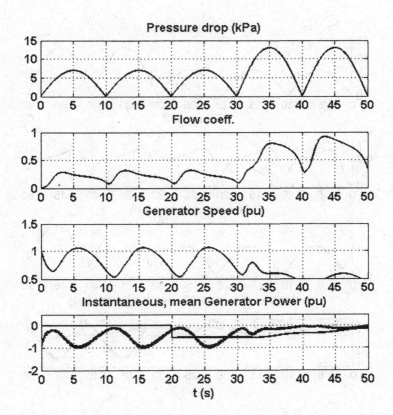

FIGURE 5.29
Processes in OWC with speed control under sharp pressure increase.

Here P_{gen_mean} is the generator power, averaged for sufficient large running window, for example, for last 10 s, the coefficient

$$K_d > 0 \text{ under } \frac{dP_{gen\_mean}}{dt} > 0, \qquad K_d = 0 \text{ under } \frac{dP_{gen\_mean}}{dt} < 0.$$

In the absence of the second term in Equation (5.52), the same picture takes place as in the previous case: with a significant pressure increase, the coefficient φ increases, the torque of the turbine and its rotational speed drop sharply and the normal mode is not restored. Figure 5.32 shows the same process as in Figure 5.30 when both terms in Equation (5.52) are taken into account. It is seen that the system functions in a normal way, and in contrast to the previous version, pressure measurement inside the turbine was not required.

The proportional integral (PI) controller of the φ value is utilized in the model **OWC2d**; the reference value is constant and equal to 0.3. The controller output is a reference for the speed controller, the output of which

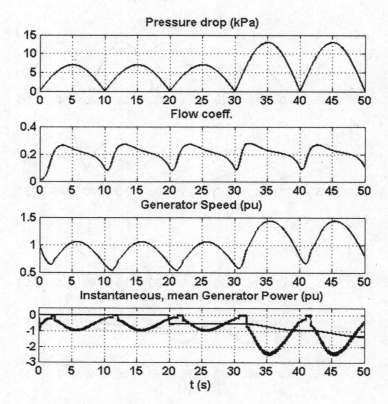

FIGURE 5.30
Processes in OWC with speed control and φ limitation.

assigns the torque of the DTC system. The process is shown in Figure 5.33, when at the times 0, 20 s, 40 s, and 60 s the values of the pressure drop amplitude were equal to 7 kPa, 13 kPa, 5 kPa, and 9 kPa, respectively. It can be seen that the mean generator power is close to the optimal values at a given pressure: 0.51 pu, 1.28 pu, 0.31 pu, and 0.74 pu, respectively. The coefficient φ remains within 0.3. The maximum possible mean power values can be found by the following way: to calculate the values of ω_t by Equation (5.41) with φ = 0.3 for the given values of the dP amplitude and afterward to use the above-given cubic expression for the optimal power; we obtain 0.54 pu, 1.38 pu, 0.33 pu, and 0.79 pu, that is, some more than the values found by simulation since in the process of controlling the values of the rotational speed and the coefficient φ somewhat deviate from the optimal values.

The VSI-Gr, related circuits, and the control system taken from the model **OWC2** are added in the model **OWC2e**. Processes in the system when the dP amplitude increases from 7 kPa to 13 kPa at t = 20 s are depicted in Figure 5.34.

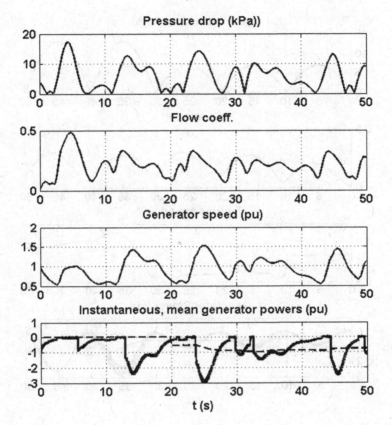

FIGURE 5.31
Processes in OWC with speed control and φ limitation under irregular waves.

Along with the considered Wells turbines, the other type turbines can be used in the OWC, for example, impulse turbines, Refs. [29–31].

These turbines have a more complex structure, but do not have a stop mode. The equations describing the behavior of these turbines are the same as the Wells turbines, but the coefficients of C_a and C_t are different. On the segment $[0 \; \varphi_{opt}]$, in which the Wells turbine usually operates, the C_t value is very small, but it increases dramatically with increasing φ. The value of C_a for an impulse turbine is smaller and has the form of a saturation curve. The shape of these curves is essentially determined by the design of the turbine. In the following, the $C_t(\varphi)$ and $C_a(\varphi)$ curves constructed using the data given in Ref. [29] are used. These curves are shown in Figure 5.35 and approximated by the expressions:

$$C_t = 0.3457(x+s)^4 - 1.9381(x+s)^3 + 3.3393\,(x+s)^2 - 0.3798(x+s) - 0.11,$$

$$(5.53)$$

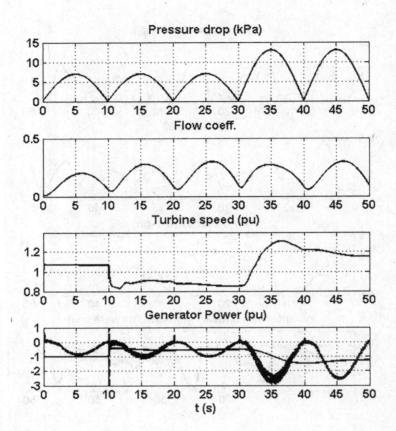

FIGURE 5.32
Processes in OWC with speed reference by Equation (5.52).

where the shift s in this case is 0.1,

$$C_a = 2.18 \times \operatorname{arctg}(4x). \tag{5.54}$$

The turbine efficiency that equal to

$$\eta = \frac{C_t}{C_a \varphi} \tag{5.55}$$

reaches maximum at $\varphi \approx 1$.

Furthermore, the turbine parameters from Ref. [30] with some alternations are used: $r = 1.5$ m, $a = 2.5434$ m^2, $J = 200$ kg m^2, and $K = 4.52$ kg/m. A comparison of these parameters with the Wells turbine parameters adopted above, as well as keeping in mind the dependence $C_a(\varphi)$, leads to the conclusion that the rotational speed of the impulse turbine is significantly less than the rotational speed of the Wells turbine. Therefore, in the model **OWC3c**, the

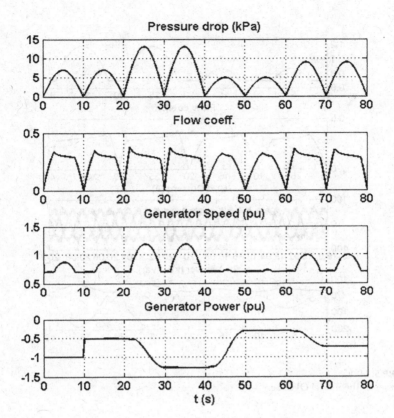

FIGURE 5.33
Processes in the model **OWC2d**.

PMSG is employed, having 20 pairs of poles and connected directly to the turbine without a gearbox. The rotor flux linkage is $\Psi_r = 0.8$. The moment of inertia of the generator with a turbine is 20 kg m². The PMSG is connected to the grid 400 V with two back-to-back connected VSIs, as shown in Figure 2.2d; VSI-Ge controls turbine rotational speed and VSI-Gr keeps the assigned DC link voltage of 900 V. The simulation of this system proves that the dependence of the maximum power on the turbine rotational speed ω_t can be approximated by the cubic parabola $P_{max} = 72\omega_t^3$. The speed controller reference quantity is computed by the given relationship using the measured power at the generator output.

In the turbine model, coefficient φ is computed by the formula that is analogues to Equation (5.50):

$$\varphi = \int K_p \left(\frac{adP}{r^2 \omega_t^2 K} - C_a(\varphi)(\varphi^2 + 1) \right) dt. \tag{5.56}$$

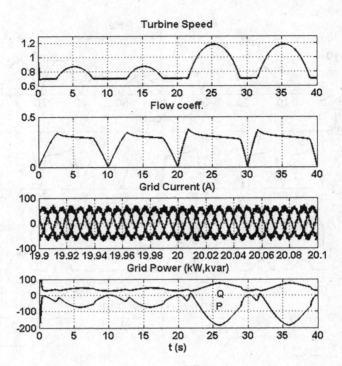

FIGURE 5.34
Processes in the model **OWC2e**.

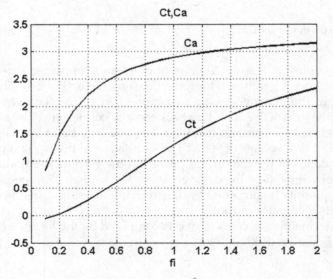

FIGURE 5.35
Coefficients $C_t(\varphi)$ and $C_a(\varphi)$ of the impulse turbine.

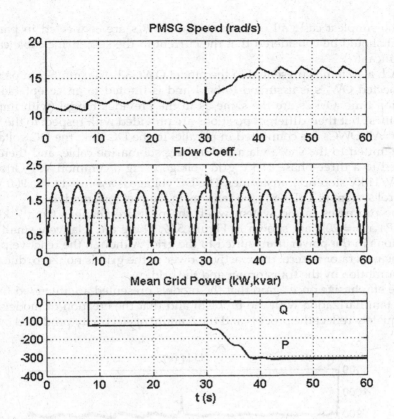

FIGURE 5.36
Processes in the model **OWC3c**.

The processes in the system with a pressure change from 7 kPa to 13 kPa at $t = 30$ s are shown in Figure 5.36. The steady-state average values of power delivered to the grid are 126 kW and 314 kW, respectively. In accordance with the above formula, the values of the rotational speed should be equal to 12 rad/s and 16.3 rad/s. The actual average values are 11.8 rad/s and 16 rad/s. Reactive power given to the grid is close to zero.

In the above text, OWCs were considered in which connection with the grid was carried out for each OWC individually. However, bearing in mind the relatively small power of individual OWCs and the limited land area, one should expect the use of OWC groups (farms), which are installed at a sufficient distance from the shore on a floating base. In this case, the problem of the expedient connection of individual OWCs and the transfer of generated energy to the shore arises. The same problems arise with the offshore installation of wind turbines; however, for OWCs there are such features as their low unit capacity and correspondingly low levels of transmitted voltage; utilization of a step-up transformer for each OWC seems impractical.

In the simplest case, all or a number of OWCs are connected in parallel, but it should be considered that the current in the submarine cable can be significant.

Such a system is presented in the model **OWC3d**. The number of parallel-connected OWCs is assumed to be 4 and is limited to an acceptable simulation time. OWCs are the same as in the previous model (with impulse turbines), but their different positions are provided with respect to the wave crest. All OWCs are connected in parallel in the DC link; the DC voltage is transmitted to the shore via a 10-km long submarine cable, and then converted to a three-phase 11 kV grid voltage using a common VSI_Gr and a 400-V/11 kV step-up transformer. The DC voltage is assumed to be 900 V, the control system of VSI_Gr is the same as in the previous model.

The processes in the system when the pressure changes from 7 kPa to 13 kPa at $t = 20$ s are shown in Figure 5.37. There are relatively small fluctuations in the power transmitted to the grid. Although the reactive power of the generator is zero, the reactive power of the grid is nonzero, due to its consumption by the transformer and the grid.

We emphasize once again that the models presented are intended for the first familiarization with the problem and that the creation of models that adequately reflect the process and the amount of electricity produced by this

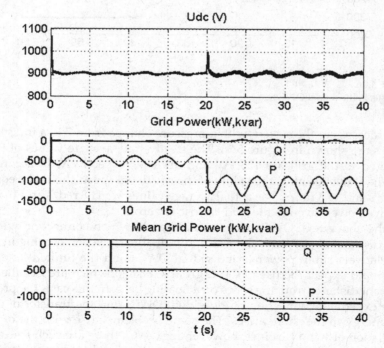

FIGURE 5.37
Processes in the model **OWC3d**.

installation requires the joint work of meteorologists, hydraulics, and electricians using the tools and means inherent in these specialties.

5.3 Ocean Tide Energy

5.3.1 Tidal Power Plant Simulation

Tidal energy is generated by the relative motion of the Earth, sun and the moon, which interact via gravitational forces. Periodic changes of water levels are due to the gravitational attraction by the sun and moon. The magnitude of the tide at a location is the result of the changing positions of the moon and sun relative to the Earth, the effects of Earth's rotation, and the local shape of the sea bottom and coastlines. Tidal power is practically inexhaustible. The more the water level height or the stronger tidal stream, the greater is the possibility to gain tidal energy for the electricity generation.

In the certain area, the tides recur regularly without significant changes, so that, unlike other renewable sources, the fabricated power can be predicted in advance.

In any point of the ocean, normally two large tides and two low tides occur each day. In general, high tides last for 12 h and 24 min. Twelve hours are due to the Earth's rotation and 24 min is due to the moon's orbit. This is the main semidiurnal cycle.

Both potential and kinetic tide energy can be converted into the electric one. In the first case, by means of barrages installed in a bay or estuary, one or more reservoirs are created to obtain a difference in water level on either side of power plant equipped with turbines. The head height that arises at that can be used to drive low head turbines.

The plant created by this way (tidal power plant) can produce an electric energy when the mass of water moves only in one direction—only at low tide or only at high tide (single-effect cycle) or when the mass of water moves in both directions (double-effect cycle). In the first case, the electricity is generated mostly at low tide, and at high tide the reservoir is filled with water through sluice gates, although systems are known when electricity is produced at high tide (e.g., the South Korean enterprise Sihwa), which is determined by the relief features. In this case, the energy is produced in two cycles per day with a duration of about 4 h each. In the second variant, reversible turbines are used, and the energy is produced during 4 four-hour cycles per day.

The energy available from barrage is dependent on the volume of water. The potential energy contained in a volume of water is

$$E = 1/2 A \rho g h^2 \; (\text{J}), \tag{5.57}$$

where h is the vertical tidal range (m), A is the horizontal area of the barrage basin (m^2), ρ is the density of water $= 1025$ kg/m^3, and g is the acceleration due to the Earth's gravity $= 9.81$ m/s^2.

The factor half is due to the fact that as the basin empty through the turbines, the hydraulic head over the dam reduces. The maximum head is only available at the moment of low water, assuming the high water level is still present in the basin, Ref. [32].

Since the produced energy is proportional to the square of the water level, the ability to run the turbines in a pump mode is used in some cases to provide an additional gain in production. The pump is used to raise the reservoir level after filling at high tide in order to increase the head when it is emptied.

The current largest tidal power plants are Rance in northwestern France and Sihwa in South Korea. The former is equipped with 24 turbines having a power of 10 MW each, the reservoir surface is 22 km^2, and the average height of the tide is 8.5 m. The electric plant employs double-effect cycle (not often) and the pump mode, Ref. [7].

The Korean tidal electrical plant is equipped with 10 turbines having a power of 25.4 MW each, the reservoir surface is 43 km^2 and the average height of the tide is 5.6 m.

The bulb-type turbines are used mostly in the modern tidal electrical plants. They are a one-piece turbine-generator and compact, made up of a Kaplan-type axial turbine, which directly drives a generator running inside a sealed housing with a bulbous profile. This type can operate in low head hydroelectric power stations reversibly as well as in the pumping mode and allow for changing the blade position. Since high-power generators are usually employed in tidal electrical plants, SGs are used as such, which are connected to the network directly or via step-up transformers. The frequency of rotation of the generators is determined by the frequency of the network and the number of pairs of poles. The excitation system regulates the excitation current in order to obtain the set value of reactive power; often the given value is zero. The turbine is equipped with additional devices: mobile blades and conical distributor that are controlled by the servomotors and are intended for turbine performance optimization and for the generator power limitation under unusual high tides. The turbine wheel rotates at a relatively low speed: 50–70 rpm, sometimes a bit more, which, among other things, is caused by the desire to reduce the mortality of the fish stock passing through the dam.

Figure 5.38 shows a possible single-effect cycle—when the generators operate at low tide. The station's parameters are close to those in Rance: 24 turbines with water flow of each one $Q = 275$ m^3/s; the minimum water head for turbine to operate is 3 m, the reservoir surface area is $S = 22$ km^2.

When water lifts, the reservoir is filled to the water level. At the instant A, the tide reaches its maximum, the gates close; the water level in the reservoir is kept at the maximum level (9 m in Figure 5.38). At low tide, an excess of the

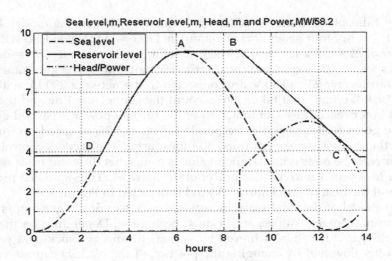

FIGURE 5.38
Single-effect cycle of the tidal power plant.

water level in the reservoir over the ocean level appears. When this excess reaches the value of 3 m, the turbines start to work (instant B). At the same time, the water level in the reservoir U decreases with the speed

$$\frac{dU}{dt} = \frac{24Q}{S}. \tag{5.58}$$

In our case

$$\frac{dU}{dt} = \frac{24 \times 275}{22 \times 10^6} = 0.3 \times 10^{-3} \text{ m/s} = 1.08 \text{ m/h}.$$

Turbines continue to work until the water head exceeds $H = 3$ m, that is, to instant C. The turbine run duration is 5 h.

The mechanical turbine power P_m may be found as

$$P_m = \eta \rho g Q H, \tag{5.59}$$

where H is the water head and η is the efficiency factor. In the case under consideration, taking $\eta = 0.877$,

$$P_m = 0.877 \times 9.81 \times 1025 \times 275 \times 24 \times H/10^6 = 58.2 \times H \text{ MW}.$$

At the instant C, the gates close and the water level in the reservoir remains at the reached level (3.8 m in Figure 5.38), until the next tidal wave at time D exceeds this level.

The tidal plant under consideration is investigated in the model **Tide_ mod1**. Twenty-four generators are simulated as one aggregated SG with the power of 24 × 11 kVA, 13,800 V. The generator has 32 pole pairs, its synchronous speed is 93.75 rpm and the inertia constant is 10 s. The generators, via the transformer 230/13.8 kV, are connected to the network 230 kV with the help of R-L circuit that takes into account the connecting line and the network inner reactances. Computation of the turbine power is carried out by above-given formulas and corresponds to Figure 5.38; to speed up simulation, the processes are performed 60 times faster. The exciters control generator reactive power; its assigned value is somewhat different from zero in order to obtain a zero reactive power of the network. The simulated process starts at the instant *B* in Figure 5.38 and is depicted in Figure 5.39.

If the level of the tidal wave increases, the power of the generators may reach unacceptable values, and it must be limited. Depending on the turbine design, this can be achieved by various means: by blocking a part of the water flow and by changing the position of the blades or guide vanes.

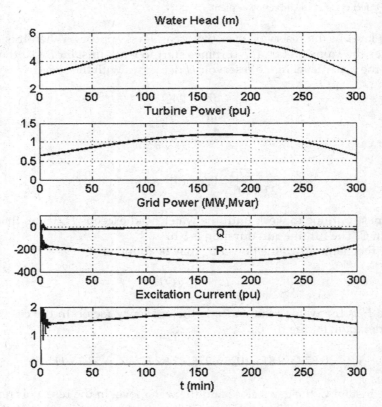

FIGURE 5.39
Processes in the model **Tide_mod1**.

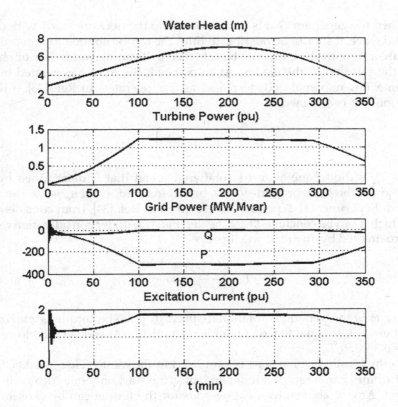

FIGURE 5.40
Processes in the model **Tide_mod2** with flow limitation.

The model **Tide_mod2** assumes a stream restriction with the help of gates. The generator power is measured; when its value exceeds 1.2 pu, the flow value is reduced by the controller **Power_lim**. The amplitude of the tidal wave is assumed to be 11 m; without limitations the power of the generators reaches 1.6 pu. As can be seen from Figure 5.40, when a restriction is introduced, this power is limited to 1.22.

If the turbines operate in both directions of water flow, and in addition, the pump mode is used, the generator must be connected to the network six times a day. Each generator connection to the network is accompanied by a certain stress for its mechanical part and accelerates its wear. Therefore, measures are taken to facilitate the start-up process. Start can be carried out from the network or using a tidal wave. In the first case, it is necessary to take measures to reduce the starting currents and the starting torque surge. This is achieved either by using a starting reactor or by a thyristor starting circuit, as in the model **HydroM3**, Ref. [3]. The same device can be used to start several generators sequentially; however, this reduces the working time of the generators, which is already limited.

When the generator that is not connected to the network starts with using a tidal wave, it should be possible to limit the torque applied to it in order to ensure a smooth start; this can be achieved by limiting water flow or changing the position of the blades. At zero speed, the torque generated by the turbine T is maximal and decreases as it accelerates. In Ref. [33], a linear relationship is assumed as

$$T = T_m \left(1 - \frac{n}{n_m} \right). \tag{5.60}$$

Here T_m is the torque at zero's rotational speed that is determined by the position of the gates or guide vanes, or turbine blades and n_m is the runaway speed. Experimental dependences are given in Ref. [34], from consideration of which it can be concluded that the dependence under consideration can be approximated by an arc of an ellipse:

$$T = T_m \sqrt{1 - \frac{1.21}{\sqrt{T_m}} (\omega - 0.3)^2}, \tag{5.61}$$

where ω is the generator rotational speed in pu. The torque is maximum T_m at $\omega = 0.3$, is equal to 0.64 under $\omega = 1$ and $T_m = 1$, and is equal to zero at $\omega_m = 0.3 + 0.9\sqrt[4]{T_m}$.

The hydraulic turbine start is simulated in the model **Tide_mod3a**. Since the turbines can be started independently, the start only one turbine is simulated. Any of above-given expressions for the torque can be chosen. The reduction of the torque due to constant and variable (proportional to the speed of rotation) losses is taken into account. To speed up the simulation, the inertial constant is halved.

Switching on the breaker connecting the generator to the network occurs when three conditions coincide: the difference between the generator and network frequencies does not exceed 2 Hz, the generator and network voltage difference (after the transformer) does not exceed 10%, the phase difference between the generator and network voltages does not exceed 0.1 rad.

Let us dwell on the last question in more detail. With the help of Phase Lock Loops (PLLs), the quantities $u_t = \sin(\omega_t t + \varphi_t)$ and $u_g = \sin(\omega_g t + \varphi_g)$ are fabricated, which have the same phases as the phase voltages of the generator and the network, respectively; and also the quantities $v_t = \cos(\omega_t t + \varphi_t)$, $v_g = \cos(\omega_g t + \varphi_g)$; and the quantity $w = u_t v_g - u_g v_t = \sin(\varphi_t - \varphi_g)$ is calculated. Since this value is zero not only when the phases are equal, but also when they are shifted by π, the value $z = v_t v_g + u_t u_g$ is computed additionally. Value $z = 1$ under $\varphi_t = \varphi_g$; the command to switch on the breaker is given when the value of z is close to 1, for example, more than 0.8. The start with implementing the dependence Equation (5.60) is depicted in Figure 5.41. To the time $t = 20$ s start is completed, and the valve is opened, the water flow increases, and the generator goes to the rated power. It is seen that the generator current

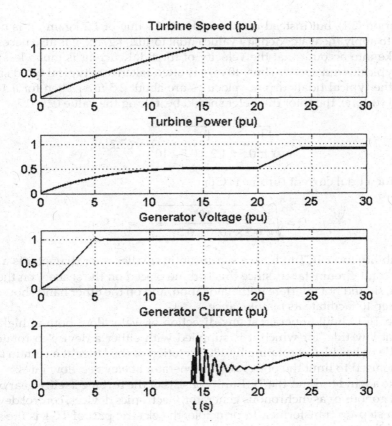

FIGURE 5.41
Start process in the model **Tide_mod3a**.

at the time of synchronization does not exceed 1.5 nominal values. Using the oscilloscope in the subsystem **Start** one can monitor the synchronization process.

When dependence Equation (5.61) is implemented, the start process runs similarly, but the generator current at the time of synchronization is greater, so that it may be necessary to change the parameters of the start circuits and exciter.

5.3.2 Tidal Current Plant Simulation

In recent years, the investigations associated with the use of the kinetic energy of tidal currents have been intensively developed. In this case, the turbines are installed on the seabed and fully immersed in water. There is a variety of such turbines, but the tendency is to use technical solutions tested in wind generators: a horizontal three-bladed turbine having a nacelle resembling the nacelle of wind generators, Ref. [35]. The turbine power is determined by

Equation (2.1), but instead of the air density value of 1.2 kg/m³, it is necessary to apply the water density value equal to 1020 kg/m³; it is also necessary to take into account that the velocity of the tidal current is much less than the typical wind speed. Thus, 10 m/s is often taken as a typical wind speed, and the typical tidal current velocities are about 2.5 m/s. Then for a 1-MW wind turbine, the rotor diameter should be, taking the value $C_p = 0.5$,

$$D = \sqrt{\frac{4}{\pi} \frac{10^6}{0.5 \times 1.2 \times 0.5 \times 10^3}} = 65.2 \text{ m},$$

and for a tidal current turbine (TCT)

$$D = \sqrt{\frac{4}{\pi} \frac{10^6}{0.5 \times 1020 \times 0.5 \times 2.5^3}} = 17.9 \text{ m}.$$

Possibility for a TCT to have a significantly smaller rotor diameter is a very important circumstance, since this turbine placed on the seabed, on the one hand, should not interfere with navigation, and on the other hand should not be deep to facilitate its maintenance.

The TCT should operate at any direction of water flow, both at high tide and at low tide, for which it is supplied with either a device to rotate the nacelle or has blades rotating by 180°; rotation of the latter at a certain angle is also used to limit the power of the generator at elevated flow rates.

Like a wind turbine, the TCT unit consists of the turbine itself, a gear box, a synchronous or asynchronous generator, electronics devices, control devices, and a step-up transformer. In principle, the electric part of TCT is the same as offshore wind generators, and the options discussed in Chapter 2 can be used for simulation of TCT. The specificity is the dependence of the power factor C_p on the parameter λ and on the angle of rotation of the blades β. Various relationships for this dependence are given in the literature. Some of them have the same form as for wind generators, but with other coefficients. So, in Ref. [36], the following relationship is used:

$$C_p = c_1 \left(\frac{c_2}{z} - c_3 \beta - c_4 \beta^{c_5} - c_6 \right) \exp\left(-\frac{c_7}{z} \right) \tag{5.62}$$

$$\frac{1}{z} = \frac{1}{\lambda + c_8 \beta} - \frac{c_9}{1 + \beta^3}. \tag{5.63}$$

At that, the vector of the coefficients c = [0.18, 85, 0.38, 0.25, 0.5, 10.2, 6.2, 0.025, −0.043], $C_{pmax} = 0.43$ and reaches at $\lambda_m = 4.25$. The C_p curves are shown in Figure 5.42. In Ref. [37] it is taken c = [0.45, 48.2, 0, 0.001, 2, 6.5, 7.2. 1, −0.001]. $C_{pmax} = 0.42$ and reaches at $\lambda_m = 3.8$.

In view of the above-mentioned analogy of TCTs and offshore wind turbines, only one model is given further: **Tide_mod6/Tide_mod6a**. As it was

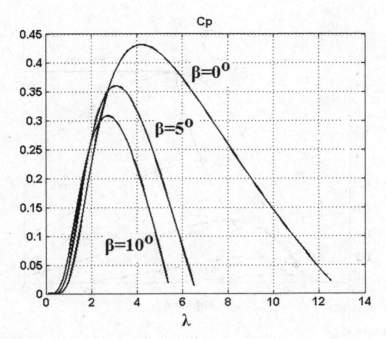

FIGURE 5.42
Dependence of C_p on λ and β for the tidal current turbine.

mentioned above, the latter model is a simplified version of the former, in which the VSI-Gr and the circuit connecting with the grid are replaced with the DC source. It is taken $D = 19.1$ m, the turbine rotates the PMSG through a speeder 1:30. The PMSG has eight pairs of poles, a nominal rotation speed of 400 rpm, and a rotor flux linkage of 2.34 Wb, so that the nominal effective value of the line voltage without load is

$$U_{leff} = 2.34 \times 8 \times \frac{\pi \times 400}{30} \frac{\sqrt{3}}{\sqrt{2}} = 960 \text{ V};$$

the DC link voltage is accepted as 2400 V. The dependencies of the turbine power on its rotational speed at various tidal current velocities are shown in Figure 5.43. It also shows the dependence of the optimal power on the rotational speed at $\beta = 0$. This curve is bounded below by the current velocity of 1 m/s and above by the maximum power of 1.7 MW at a current velocity of 3 m/s. The diagram of the turbine model is shown in Figure 5.44.

The torque of the turbine can be defined as

$$T_m = P_m / \omega_r = 0.5\rho \frac{\pi D^2}{4} C_p \frac{\omega_r^2 D^3}{8\lambda^3} = 0.5\rho \frac{\pi D^5}{32} C_p \frac{\omega_r^2}{\lambda^3} = 1.27 \times 10^8 \frac{C_p \omega_r^2}{\lambda^3}, \quad (5.64)$$

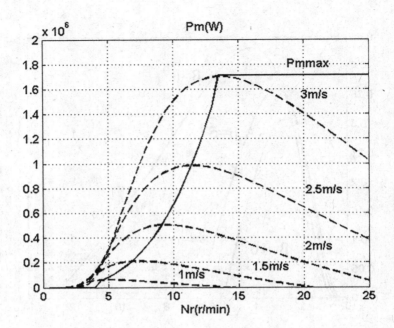

FIGURE 5.43
Dependence of the TCT power on its rotational speed at various tidal current velocities.

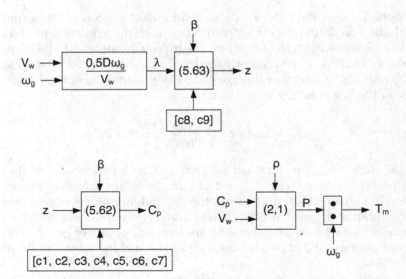

FIGURE 5.44
Block diagram of the TCT model.

where ω_r is the turbine rotational speed. When the generator power exceeds 1.74 MW, the circuits affecting blade position are activated, which consist of PI controller, the holder, and the limiter of the speed of the blade position change β, $\beta \leq 20°$. The inverter VSI-Ge controls the i_q component of the stator current, the reference for $i_d = 0$. There are two options for i_q^* current reference: in the first case, the value is determined by the output of the speed controller; its preset value is proportional to the current velocity V_w with the coefficient of proportionality that is determined by λ_m. In the second case, the i_q^* value is proportional to the value of the optimal turbine torque at the known actual speed of its rotation; as shown in Chapter 2, maximum power point tracking (MPPT) is provided. Since, based on Equation (5.64), the optimal generator torque at the given rotational speed ω_g is equal to

$$T_g = \frac{1.27 \times 10^8 \times 0.43}{i^3 \times 4.25^3} \omega_g^2 = 26.3\omega_g^2,$$

then

$$i_g^* = \frac{26.3}{8 \times 2.34 \times 1.5} \omega_g^2 = 0.94\omega_g^2.$$

It is assumed that the tidal current velocity changes sinusoidal with time; the velocity at high tide is greater than the velocity at low tide, Ref. [35]. The maximum velocity at high tide is assumed to be 3.5 m/s, and at low tide it is 2.5 m/s. To reduce the simulation time, the wave period is reduced by 60 times.

At the current velocity of 0.9 m/s, power generation stops. There are various scenarios of system behavior in this case. It is assumed in the model that in this case a *Stop* signal is fabricated, by which a mechanical brake is applied, the speed regulator is set to zero, the fabrication of gate pulses to the voltage inverter stops. There are possible (but not considered) alternatives with the installation of blades in the neutral position and with the continued rotation of the rotor at low speed.

In the model **Tide_mod6a**, to reduce the simulation time, the network inverter is not simulated; it is assumed that it maintains the voltage in the DC link equal to the specified value. Figures 5.45 and 5.46 show the processes in the system for both variants of the set point i_q^* determination, it is clear that the results are close, but the second option does not require accurate and continuous measurement of the tidal current velocity. When its value exceeds 3.5 m/s, the unit for blade position adjustment comes into operation, limiting the power produced by the generator. Power load accurately follows the optimal power curve. So, for the current velocities of 2 m/s and 2.5 m/s, the maximum power values are 0.51 MW and 1.01 MW, respectively, while the simulation results for options 1 and 2 give 0.504 MW, 0.985 MW and 0.503 MW, 0.987 MW, respectively (it is turbine powers measured at low tide).

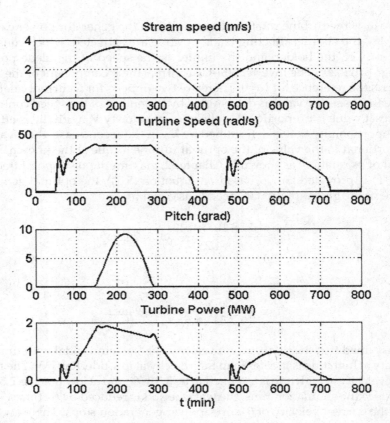

FIGURE 5.45
Processes in the model **Tide_mod6a** with i_q^* that is proportional to V_w.

The model **Tide_mod6** is a complete system. The network inverter VSI-Gr keeps the constant voltage in the DC link U_{dc} by acting on the network current component i_q and adjusts the amount of reactive power delivered to the network by acting on the component i_d. The energy produced by the generator, through the VSI-Gr, step-up transformer 1/35 kV and a 20-km transmission line, is transmitted to the network.

The processes in the system are shown in Figure 5.47. Since the simulation in this system runs slowly, only part of the cycle including stop mode is simulated. At the low current velocity of 0.9 m/s, the electrical energy production stops, but the VSI-Gr remains connected to the grid, practically without consuming current. Of course, the option of disconnecting the VSI-Gr from the network is possible. With an increase in the current velocity, power generation is resumed and at the end of the simulated time, at the current velocity of 3.04 m/s, the β control unit enters into operation, the power transmitted to the network is limited to 1.7 MW, the reactive power is maintained close to zero.

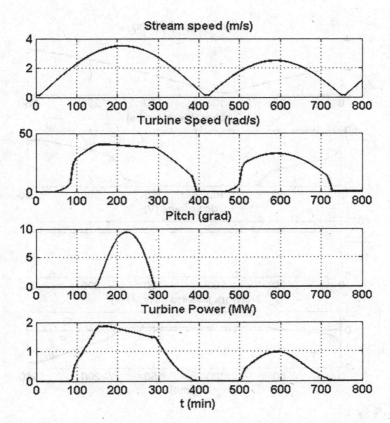

FIGURE 5.46
Processes in the model **Tide_mod6a** with i_q^* that is proportional to optimal turbine torque.

5.4 Pumped Storage Hydropower

The power management and grid stabilization are of great interest. This problem is complicated by an increase in renewable source penetration, with their intermittency power production. To overcome this problem, efficient and economic technologies to store large amounts of electrical energy are needed. Pump-storage power stations are the very efficient technology to solve this problem.

A typical pump-storage power station consists of two interconnected water reservoirs (an upper and a lower), tunnel or waterway that convey water from one reservoir to another, and a powerhouse with a pump/turbine, a motor/generator, and an appropriate electrical equipment. When there is low power demand and electricity is inexpensive (which typically occurs overnight), the water is pumped from the lower reservoir to the upper reservoir.

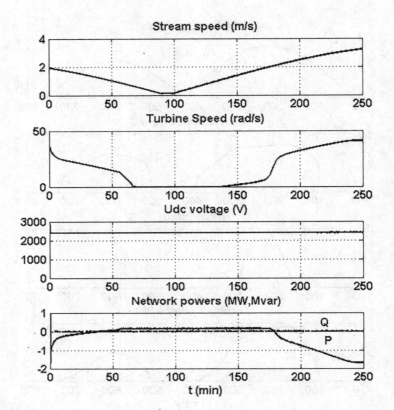

FIGURE 5.47
Processes in the model **Tide_mod6**.

Water is stored in the upper reservoir and is released during peak demand periods, delivering more valuable electricity to the grid. In the first case, an electrical machine works as a motor, and in the latter case, as a generator.

The constructions exist, in which a turbine and a pump are the different elements located on the same shaft with the generator/motor (so-called ternary configuration), and two-unit constructions, in which the runner operates in both modes: both a turbine and a pump. In the second construction, the rotational direction is reversed for mode change; it takes a perceptible time and is a stress for the equipment. In the ternary configuration, a mode change is carried out without reversing that saves time.

The efficiency of a turbine/pump with the constant rotational speed decreases when the running conditions differ from the rated ones. Fortunately, although the powers of the generators can reach tens and hundreds of MWs, state-of-the-art of power electronics makes it possible to use its components for the rotational speed control, and thereby to optimize the operation of the unit under shifting conditions. Besides, the process of a speed reversal is more favorable that promotes an employment of a two-unit configuration.

Both the synchronous machines (SMs) with an excitation winding on the rotor and the doubly fed induction machines (DFIMs) are utilized in the modern pump-storage plants. The DFIM is more attractive in high-power applications because the converter itself can be considerably smaller, Ref. [38]. But hydropower systems with DFIMs have a number of drawbacks, including existence of slip rings, limited range of the speed control, and more complicated start-up procedure. The DFIM power in present hydropower systems reaches 250 MW and more in one unit.

The SMs in hydropower plants need the full-rated converter, which could be very expensive and requires a large room. These problems limit the application of SMs for hydropower plants with power ratings not more than 100 MW. But its superior performance over the DFIM and progress in power electronics may provide the opportunity for the SM to be the preference option even in higher power sets, Ref. [39].

In the regulated pump/turbine sets, there are two variables to control: the flow rate Q with the help of guide-vanes, as in the uncontrolled sets, and the turbine rotational speed ω_r. One of them is responsible for production of the reference power, and another for the optimal efficiency. The used pump/turbine model has to take into account this feature; so, this model turns out to be more complicated. Therefore, a number of simplified models are considered in the published works. So, in Ref. [40], the turbine model developed in Ref. [41] and employed in Ref. [1] (see Ref. [2] as well) is applied for turbine and pump modes. The speed variations are not considered in the model; it is supposed that the reference speed is defined by relationships supplied by the manufacturer.

The model used in Ref. [42] has some additional features. Its relationships are

$$T_w \frac{dQ}{dt} = H_s - H_d - dH,$$ (5.65)

where H_s is the static head from the upper reservoir to the pump-turbine inlet, H_d is the dynamic head, dH is the frictional head loss that is equal to $f_p \times \text{sign}(Q)Q^2$, and T_w is the water starting time that is a few seconds usually; all the quantities are in pu.

In the turbine mode, dynamic head is defined as

$$H_d = \left(\frac{Q}{K_g g}\right)^2,$$ (5.66)

where K_g is the gate opening factor that is equal to $1/(g_{max} - g_{min})$, they are the limits imposed on the gate opening, and g is the gate position.

In the pumping mode, dynamic head H_d is determined as

$$H_d = a\omega_r^2 - b\omega_r Q - cQ^2,$$ (5.67)

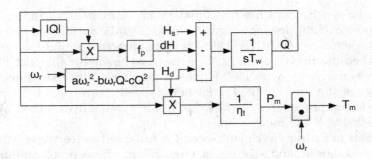

FIGURE 5.48
Block diagram of the model of the turbine/pump in pumping mode.

where a, b, c are the coefficients, which are taken as $a = 1.4$, $b = 0.2$, and $c = 0.3$. The mechanical power in the turbine mode is

$$P = H_d (Q - dQ),\qquad(5.68)$$

where dQ is the no-load flow. All the quantities are in pu. In the pumping mode $dQ = 0$, but the efficiency η_p is taken into account.

A block diagram in the pumping mode, for the case of full-open gate (in order to decrease loss), is depicted in Figure 5.48. The power and flow rates as the functions of the rotational speed for different static heads H_s are displayed in Figure 5.49. One can see, for example, that at $\omega_r \approx 1.1$ the values of

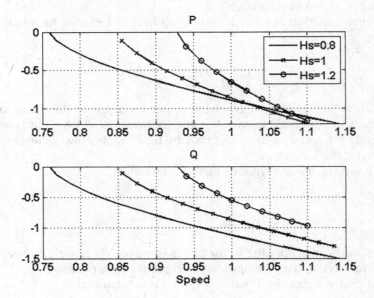

FIGURE 5.49
Dependencies of the power and flow rate on the rotational speed and water head.

the power are nearly the same, but for the case of the lower head the flow rate is appreciably more; it means that for the same time, the higher water volume can be transported up from the lower basin to the upper one.

The model **Pump1** models a pumped storage hydro-plant equipped with the SG, which is direct connected to the grid. Therefore, SG rotational speed is constant. The SG power is 200 MVA at the voltage of 13,800 V, it is connected to the grid with a voltage of 500 kV by means of the transformer 500/13.8 kV. The block diagram of the turbine control system is displayed in Figure 5.50. A number of signals sum up at the input of the PI controller. Because the steady-state speed is constant, the speed difference has a stabilizing action only. The main loop is the gate position control with aim to provide the reference power in the turbine mode. Because in the steady state $H_s = H_d + dH$, it follows from Equations (5.66) and (5.68) under $f_p = 0$, $\eta_p = 0$

$$g = \frac{P + dQH_s}{K_g H_s^{1.5}}.$$
(5.69)

The additional difference $P_{ref} - P_e$, where P_e is the SG power corrected the reference gate position.

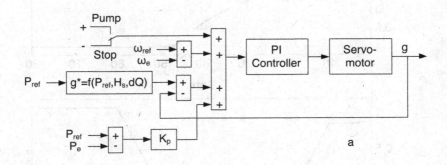

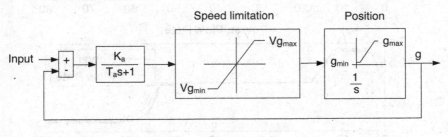

FIGURE 5.50
Block diagram of the hydraulic turbine control system. (a) Gate control and (b) servomotor control.

Initially, the process is simulated when the plant works in the turbine mode with $P_{ref} = 0.8$; at $t = 10$ s the command appears to stop this mode. When the gate closes, the flow stops. At $t = 20$ s the command is generated for the SG to change the direction of rotation. For this, the phase sequence of the supply voltage alternates. For simplicity and simulation speedup, the steps are not taken to decrease the current inrush. The SG operates as an asynchronous motor, because SG excitation is disconnected; it recovers when the SG reaches nearly synchronous speed. At $t = 50$ s the pumping mode begins, the gate opens fully, the water flows in the opposite direction (Figure 5.51).

The SG with variable speed is considered in the models **Pump2, Pump21, Pump22**. The SG has the same parameters as in the model **Pump1**, but its power decreases to 50 MVA. The hydroturbine control system sets the gate position in accordance with the reference power, relationship (5.69). From the

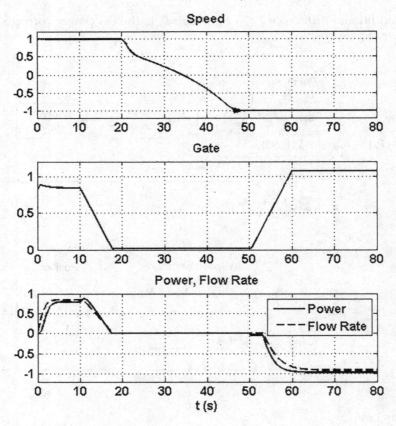

FIGURE 5.51
Transition from turbine to pump mode for SG with constant rotational speed.

results of computations given in Figure 5.49, one can conclude that the dependence of ω_{ref} on P_{ref} can be approximated as

$$\omega_{ref} = 0.0715\frac{P_{ref}}{H_s} + 0.846\sqrt{H_s}\left[1 + 0.15\left(\frac{P_{ref}}{H_s}\right)^2\right]. \tag{5.70}$$

However, under $0.35 < P_{ref} < 1.05$, the simpler expression may be used:

$$\omega_{ref} = 0.79 + 0.25\frac{P_{ref} + 1.8(H_s - 1)H_s}{H_s}. \tag{5.71}$$

The dependences of Equations (5.70) and (5.71) are plotted in Figure 5.52. It is seen that the former coincides with the plots in Figure 5.49, and the latter can be utilized, in certain cases, in the above-pointed range as a reference for the speed controller, but at this, it is necessary to have in mind that the dependence of $P_{ref} = f(\omega_{ref})$ is rather steep, and, under nominal conditions, the 1% speed change brings 4.2% power change.

The model **Pump2** (**Pump2N** for R2019a) is meant for quick investigation of the system main special features. The SG is connected with the two-level VSI (VSI-Ge) that controls SG rotational speed. The grid VSI (VSI-Gr) is not simulated; it is supposed that the DC voltage is kept constant and that is 25 kV.

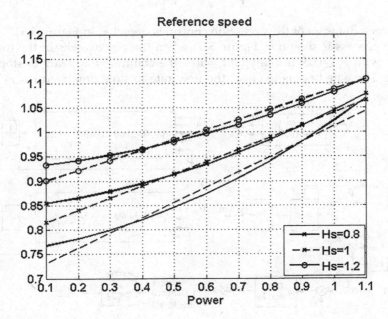

FIGURE 5.52
Accurate and approximate dependencies of the reference rotational speed from the wanted power.

The SG control system keeps SG magnetizing flux linkage Ψ_m affecting its excitation voltage. This control system has the inner controller that controls the excitation current I_{fd}. The exciter is modeled as a lag element. The flux linkage Ψ_m is computed by the following way.

Calculations are executed in the stationary axes α/β. The components $\Psi_{m\alpha}$ and $\Psi_{m\beta}$ are

$$\Psi_{m\alpha} = \Psi_{s\alpha} - L_{sl}I_{s\alpha}, \qquad \Psi_{m\beta} = \Psi_{s\beta} - L_{sl}I_{s\beta}, \tag{5.72}$$

where $\Psi_{s\alpha}$ and $\Psi_{s\beta}$ are the components of the stator flux linkage, $I_{s\alpha}$ and $I_{s\beta}$ are the same for the stator current, and L_{sl} is the stator leakage inductance. The components $\Psi_{s\alpha}$ and $\Psi_{s\beta}$ are computed as, Ref. [43],

$$\begin{aligned} \Psi_{s\alpha} &= \int (U_{s\alpha} - R_s I_{s\alpha} + \lambda U_{s\beta} - \lambda\omega_s \Psi_{s\alpha})dt \\ \Psi_{s\beta} &= \int (U_{s\beta} - R_s I_{s\beta} - \lambda U_{s\alpha} - \lambda\omega_s \Psi_{s\beta})dt \end{aligned} \tag{5.73}$$

Here $U_{s\alpha}$ and $U_{s\beta}$ are the components of the stator voltage, ω_s is the stator voltage frequency, and the constant λ changes its sign together with the sign of ω_s. Then

$$\Psi_m = \sqrt{\Psi_{m\alpha}^2 + \Psi_{m\beta}^2}. \tag{5.74}$$

The block diagram of the excitation control is depicted in Figure 5.53.

The process is shown in Figure 5.54 when the plant works in the turbine mode with $P_{ref} = 0.9$ initially; at $t = 10$ s the command appears to stop this mode. The gate begins to close, the flow rate to slow down. At $t = 14$ s the

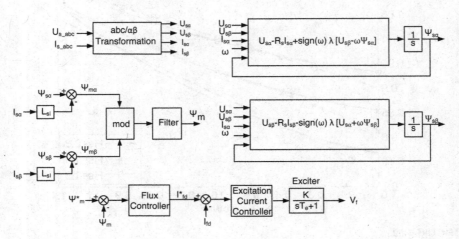

FIGURE 5.53
Block diagram of the excitation control for the pump/turbine with speed control.

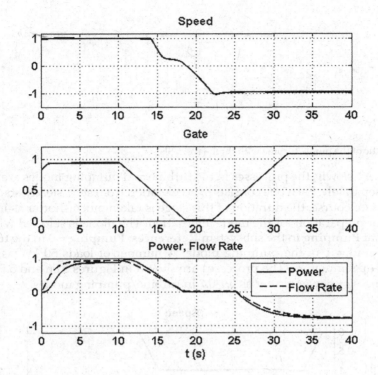

FIGURE 5.54
Processes in the model **Pump2**.

command is generated for the SG to change the direction of rotation. At about $t = 20$ s the flow stops completely, the SG rotates no-load in the opposite direction. At $t = 22$ s a pumping begins with $P_{ref} = 0.7$, the gate is opening, a water flow appears. The mechanical power reaches 0.77, owing to the losses (efficiency $< 1, f_p > 0$) and an inaccuracy of relationship (5.71).

The model **Pump21** (**Pump21N** for R2019a) is the model **Pump2**, in which two-level VSI is replaced with two three-level VSIs connected in parallel; they are connected with the secondary winding of the transformer with the nominal voltage of 5 kV; the primary winding with the voltage of 13,800 V is connected to the SG stator. The DC voltage is taken as 10 kV. The processes in this model are the same practically as in the model **Pump2**, but the higher harmonics in the SG stator current are much less: at $t = 9$ s THD $\approx 5\%$, whereas THD $\approx 11\%$ at this time in the model **Pump2**.

The model **Pump22** (**Pump22N** for R2019a) is the complete model with three-level grid VSIs. The schematic of the main circuits is depicted in Figure 5.55. The three-level VSIs are connected to the transformer different windings with the voltage of 5 kV. The inverter DC terminals are connected in parallel. The DC voltage is taken as 10 kV. Because simulation in this model

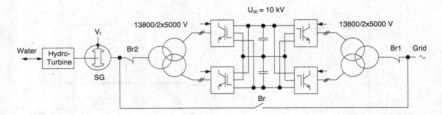

FIGURE 5.55
Schematic of the pump/turbine with three-level VSIs.

runs very slowly, the processes in the turbine and pumping modes are simulated separately. Each one consists of acceleration from stop, steady-state conditions (of course, the duration of this stage is taken much shorter as in fact), stop for preparation for the mode alternation. The mode is selected with the constant **Pumping** in the subsystem **References: Pumping** = 0 in the turbine mode, and = 1 in the pumping mode. A number of loads 50 kW assist to speed up simulation. The processes are shown in Figures 5.56 and 5.57. The references of power are set as 0.9/0.7 in turbine/pumping modes.

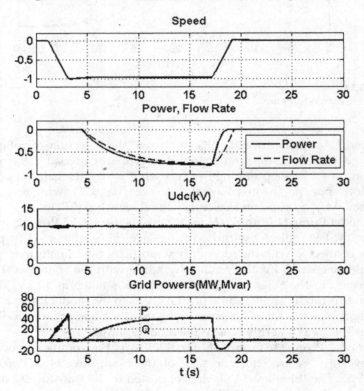

FIGURE 5.56
Processes in the model **Pump22** in pump mode.

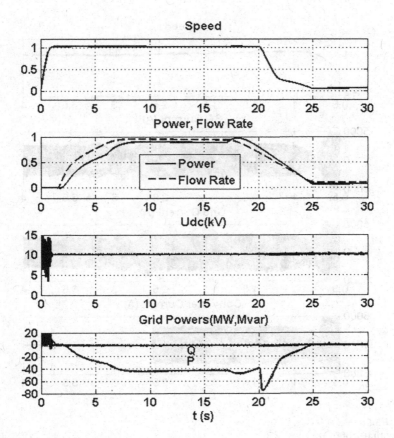

FIGURE 5.57
Processes in the model **Pump22** in turbine mode.

Bypass switch is shown in Figure 5.55. When turbine operates under the conditions close to the optimal ones for a given turbine, the loss of efficiency owing to deviation of the rotational speed from the synchronous one can be less than the loss in the converters; therefore, direct connection of the SG with the grid turns out reasonable. Besides, it makes possible to service the converter unit, while the power plant remains in operation, Ref. [39]. The converter is utilized for SG acceleration to the synchronous speed and synchronization with the grid. Because it was found that the presence of the power switches in the model slows down the simulation essentially, the separate model **Pump221 (Pump221N** for R2019a) is developed. Some changes in the subsystems **References** and **Turbine** are made to speed up simulation ($\omega_{ref} = 1$, the time constant T_w is decreased 10-fold). It is worth noting that the blocks **Ideal Switch** are used for the model of bypass switches because it was found that the simulation runs faster than with the block **Three-Phase Breaker**.

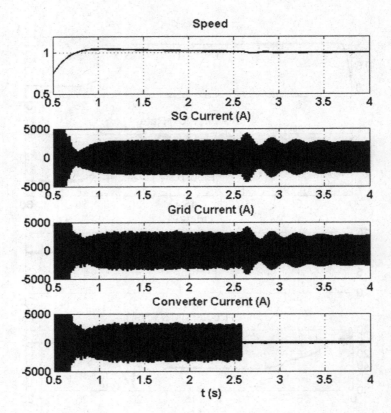

FIGURE 5.58
Start with use of the bypass switch.

For synchronization, the subsystem **Start** from the model **Tide_mod3a** is used with some modification: instead of the comparison of the frequencies, the comparison of the rotational speed with the synchronous one (1 in pu) is utilized; besides, instead of the SG voltage, its flux linkage is used. The process of acceleration and synchronization is shown in Figure 5.58.

Owing to the low switching frequency (600 Hz), the output converter current is distorted essentially, and the filter for connection with the grid is indispensable. Because an insert of the filter in the model **Pump2** slows down simulation, the separate model **Pump222** (**Pump222N** for R2019a) is employed for modeling of the system with the filter; the steps are made to speed up simulation, as made in the previous model. The filter is placed between the converter and the grid; T-type L_c–RC–L_g filter with $L_c = L_g$ is used, Ref. [44]. With the selected parameters, total harmonic distortion (THD) lessens from ~9% for the converter output current to ~3% for the grid input current (Figure 5.59). The reader can experiment with the other parameters and diagrams of the filter.

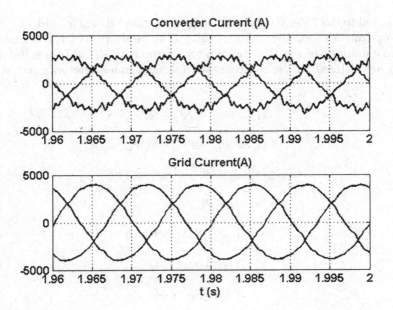

FIGURE 5.59
Grid current distortion decrease by the *L–C–L* filter.

As it was mentioned earlier, employment of the DFIG gives a possibility to work with the converters of considerably less power. Two back-to-back connected VSIs are set in the rotor circuits, as in the model **Wind_DFIG_1N**. Before to be connected to the grid, the turbine with the DFIG has to be speeded up to the rotational speed close to the synchronous one. This is achieved by powering the DFIG rotor when the stator windings are shorted. The difficulty of this operation is that the power and, consequently, the maximum voltage of the inverter connected with the rotor is much less than the nominal stator voltage equal to the supply voltage, so the DFIG must be accelerated with a significantly weakened flux, that is, with a small value of the torque. Fortunately, the resulting total mechanical losses during start-up are not more than 2.5%, Ref. [45], so the start can be realized.

After the DFIG reaches the speed close to the synchronous one, the stator windings are disconnected and the inverter gate pulses are stopped for a short time, the machine is demagnetized. After this, the rotor voltage is produced with the slip frequency; the conditions for synchronization are formed. After synchronization, the load can be given and the normal work starts.

The DFIG can be accelerated in the V/f mode or with stator flux-oriented control, Refs. [45, 46]. The latter has better characteristics, but it is more complex; only the first is applied in subsequent models. Three-level inverters are used for VSI-Ge and VSI-Gr in the model **Pump23**. The DFIG has the power of 230 MVA at the voltage of 13,800 V, 50 Hz, and nine pairs of poles. Its stator is connected to the 230-kV network using the breaker **Br** via a 230-kV/13,800-V step-up transformer. The stator windings can be shorted by the **Br1** breaker.

The equations of the DFIG given, for example, in Ref. [47] (see Ref. [3] as well), are utilized for the control system design. In *q-d* reference frame, they are

$$U_{qs} = R_s i_{qs} + s\Psi_{qs} + \omega_s \Psi_{ds} \tag{5.75}$$

$$U_{ds} = R_s i_{ds} + s\Psi_{ds} - \omega_s \Psi_{qs} \tag{5.76}$$

$$U_{qr} = R_r i_{qr} + s\Psi_{qr} + (\omega_s - \omega_r)\Psi_{dr} \tag{5.77}$$

$$U_{dr} = R_r i_{dr} + s\Psi_{dr} - (\omega_s - \omega_r)\Psi_{qr} \tag{5.78}$$

$$T_e = 1.5 Z_p (\Psi_{ds} i_{qs} - \Psi_{qs} i_{ds}) \tag{5.79}$$

$$\Psi_{qs} = L_s i_{qs} + L_m i_{qr} \tag{5.80}$$

$$\Psi_{ds} = L_s i_{ds} + L_m i_{dr} \tag{5.81}$$

$$\Psi_{qr} = L_r i_{qr} + L_m i_{qs} \tag{5.82}$$

$$\Psi_{dr} = L_r i_{dr} + L_m i_{ds}. \tag{5.83}$$

Here T_e is electromagnetic torque, $L_s = L_{ls} + L_m$, L_{ls} is the stator leakage inductance and L_m is the magnetizing inductance, $L_r = L_{lr} + L_m$. L_{lr} is the rotor leakage inductance, U_{ds}, U_{qs}, U_{dr}, U_{qr} are the stator and rotor voltage components, respectively, Ψ_{ds}, Ψ_{qs}, Ψ_{dr}, Ψ_{qr} are the stator and rotor flux linkage components, respectively, ω_s is the grid voltage angular velocity, ω_r is the rotor rotational speed (electrical), and *s* is the differentiation symbol.

The stator voltage-oriented control system is used for VSI-Ge. The grid voltage space vector amplitude is U_{ds}, at this, $U_{qs} = 0$. The block PLL measures the angle of this vector θ_s. Moreover, the rotor angle θ_r is supposed known (measured). By means of the Park transformation with the transformation angle $\theta_\Delta = \theta_s - \theta_r$, the rotor current components i_{dr} and i_{qr} are calculated, which are the feedback quantities for the proper controllers.

If to neglect the stator winding resistance, then $\Psi_{ds} = 0$, and from Equation (5.81) it follows

$$i_{dr} = -\frac{L_s}{L_m} i_{ds}. \tag{5.84}$$

The torque in pu taking into account Equations (5.76), (5.79), and (5.84) is

$$T_e = -\Psi_{qs} i_{ds} = -\frac{L_m}{L_s} U_{ds} i_{dr}. \tag{5.85}$$

Then, with the given torque T_e^*, the reference for current i_{dr} is

$$i_{dr}^* = -\frac{L_s}{L_m}\frac{T_e^*}{U_{ds}}. \tag{5.86}$$

From the expression for the reactive power

$$Q = U_{qs}i_{ds} - U_{ds}i_{qs}, \tag{5.87}$$

it follows

$$i_{qs} = -\frac{Q}{U_{ds}}, \tag{5.88}$$

then, taking Equation (5.76) into account, relationship (5.80) may be written in pu as

$$-U_{ds} = -\frac{L_sQ}{U_{ds}} + L_m i_{qr}. \tag{5.89}$$

Therefore, the wanted value of the current i_{qr} may be written as

$$i_{qr}^* = -\frac{U_{ds}}{L_m} + \frac{L_s}{L_m}\frac{Q}{U_{ds}}. \tag{5.90}$$

The outputs of the current controllers, with the help of the reverse Park transformation, are transformed into the three-phase signal that controls the VSI-Ge pulse-width modulation (PWM). The control system block diagram is depicted in Figure 5.60. It is taken for simplicity $Q = 0$.

At the beginning of acceleration, **Br1** is closed and **Br** is open. The PWM input is determined by a three-phase sinusoidal signal generator, the amplitude and frequency of which grow in time, the switch **Sw2** is in the upper position. When the DFIG reaches a speed close to synchronous, the switch **Sw2** moves to the lower position; since **Sw1** is in the upper position, the PWM input is zero. After some time, the signal *St2* appears, and the breaker **Br1** opens, the stator current disappears and the DFIG is demagnetized. Then **Sw1** closes, and a three-phase voltage with a slip frequency is supplied to the rotor, thereby creating conditions for synchronization that are the same as in the model **Tide_mod3a**: the expected DFIG supply voltage and the real voltage of its stator are compared in amplitude, frequency, and phase. The schematic of the unit for synchronization is basically the same as in the mentioned model. When synchronization occurs, the signal *Str* appears, the breaker **Br** closes, and the conditions for the DFIG load are created. Some circuits intended for smooth transition from one structure to another are not shown in Figure 5.60. For example, a multiplier is placed at the PWM input, its first input is the modulation waveform, and the second input changes smoothly from 0 to 1 under switching over in the system.

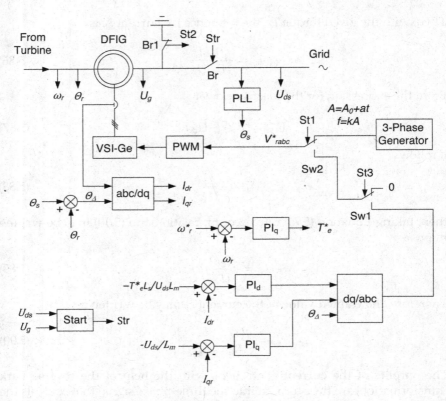

FIGURE 5.60
Block diagram control system of the pump/turbine with DFIG.

As for the VSI-Gr control system, it is the same as the previous models (see, e.g., **Pump22**): DC link voltage and the appropriate reactive power are controlled.

The process of acceleration in the pump mode is simulated in the model **Pump23**. Unlike previous models, the pump mode corresponds to $\omega_r > 0$. Signals *St1*, *St2*, *St3*, and *Str* are fabricated at $t = 30$ s, 31 s, 32 s, and 32.3 s, respectively, a load of 0.9 nominal is applied at $t = 35$ s. The process is shown in Figure 5.61.

Models **Pump24** and **Pump242** are intended to study quasi-steady-state modes. That is why some simplifications are made in the turbine model. Moreover, the turbine in the turbine and pump modes is simulated separately using the above-given relationships. The mode is selected by the *Rev* constant: in the pump mode $Rev = 0$, in the turbine mode $Rev = 1$. To change the direction of rotation, the sequence of the phases B and C is changed using the subsystem **Reverse**. Note that when changing the simulation mode, it is necessary to change the initial slip setting in the DFIG dialog box: 0.05 in the

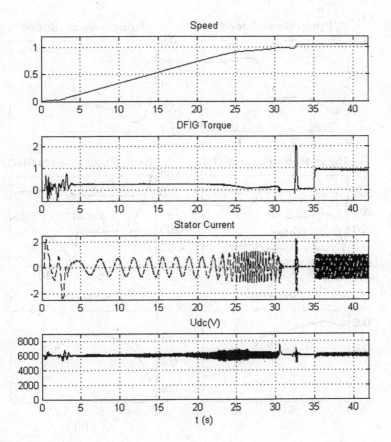

FIGURE 5.61
Process of speed up with DFIG.

pump mode and 1.95 in the turbine mode; recall that in the pump mode, the DFIG rotation speed is plus.

It is assumed in the first model that the voltage in the DC link is kept with sufficient accuracy in the steady state, so that the processes in the inverter connected to the network can be neglected, and it can be taken $U_{dc} = 6$ kV. Processes are shown in Figure 5.62, when in the pump mode the power rises from 0.6 to 1 pu at $t = 7$ s, and in the turbine mode it decreases from 1 to 0.6 at $t = 7$ s. Initial sections (up to about 3 s) can be ignored. In the first case, the power consumed from the network increases from 170 MW to 240 MW, and the slip varies from 6% to –4%, so that the power transferred to the rotor should vary from 10 MW to –9.6 MW, which corresponds to the first axis of the **DFIG Powers** oscilloscope. The flow rate increases from 0.66 to 1.02. In the turbine mode, the power delivered in the network decreases from 210 MW to 135 MW, and the slip changes from –4% to 6%, so that the power

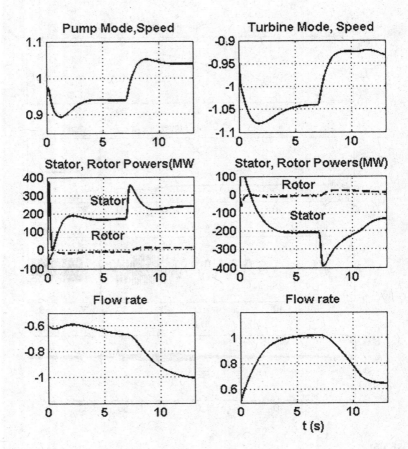

FIGURE 5.62
Processes in the model **Pump24**.

transferred to the rotor has to change from 8.4 MW to –8.1 MW, which also approximately correspond to the data on the first axis of the **DFIG Powers** oscilloscope. The flow rate decreases from 1.02 to 0.65.

In the **Pump242** model, VSI-Gr and the transformer are added to the rotor circuit, as well as additional devices (**Br, Switch**), which enable to approach the steady state that is established at $t \approx 2$ s; processes at $t < 2$ s can be ignored. If to carry out simulation, one can see that the main processes in the model are very close to those in the previous model with the same impacts; at the same time, simulation in this model is performed 2–2.5 times slower. Figure 5.63 shows changes in the rotational speed and power of the grid, the stator, and the rotor when the power setting changes from 0.6 pu to 1 pu at $t \approx 8$ s in the pump mode.

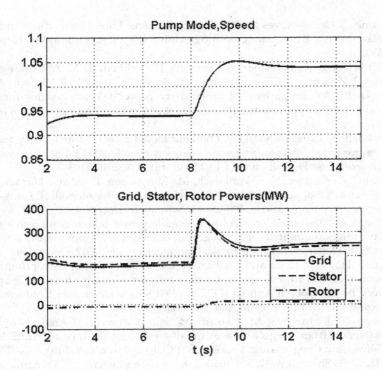

FIGURE 5.63
Processes in the model **Pump242**.

References

1. MathWorks, Simscape™ Electrical™, User's Guide (Specialized Power Systems). MathWorks, Natick, MA, 1998–2019.
2. Perelmuter, V. M. Electrotechnical Systems, Simulation with Simulink and SimPowerSystems. CRC Press, Boca Raton, FL, 2013.
3. Perelmuter, V. M. Renewable Energy Systems, Simulation with Simulink and SimPowerSystems. CRC Press, Boca Raton, FL, 2017.
4. Borkowski, D., Laboratory model of small hydropower plant with variable speed operation. Zeszyty problemowe: Maszyny Elektryczne, 3(100), 2013, 27–32.
5. Brekken, T. K. A., von Jouanne, A., and Han, H.Y. Ocean wave energy overview and research at Oregon State University. Power Electronics and Machines in Wind Applications Conference, Lincoln, NE, 2009.
6. Engineering Committee on Oceanic Resources. Wave Energy Conversion. Elsevier Ocean Engineering Book Series, vol. 6, Elsevier, Oxford, UK, 2003.
7. Multon, B. (ed.). Marine Renewable Energy Handbook. John Wiley & Sons, Inc., Hoboken, NJ, 2012.

8. Falnes, J. Ocean Waves and Oscillating Systems: Linear Interactions Including Wave-energy Extraction. Cambridge University Press, Cambridge, New York, 2002.

9. Rhinefrank, K., Schacher, A., Prudell, J., Brekken, T. K. A., Stillinger, C., Yen, J. Z., Ernst, S. G.,von Jouanne, A., Amon, E., Paasch, R., Brown, A., and Yokochi, A. Comparison of direct-drive power takeoff systems for ocean wave energy applications. IEEE Journal of Oceanic Engineering, 37(1), 2012, January, 35–44.

10. Pakdelian, S., and Toliyat, H. A. Dynamic modeling of the trans-rotary magnetic gear for the point-absorbing wave energy conversion. IEEE Energy Conversion Congress and Exposition (ECCE), Pittsburgh, PA, 2014, 3163–3170.

11. Holm, R. K., Berg, N. I., Walkusch, M., Rasmussen, P. O., and Hansen, R. H. Design of a magnetic lead screw for wave energy conversion. IEEE Transactions on Industry Applications, 49(6), 2013, November/December, 2699–2708.

12. Cargo, C. Design and control of hydraulic power take-offs for wave energy Converters. Submitted for the Degree of Doctor of Philosophy of the University of Bath, Department of Mechanical Engineering, UK, 2012, December.

13. Luan, H., Onar, O. C., and Khaligh, A. Dynamic modeling and optimum load control of a PM linear generator for ocean wave energy harvesting application. Twenty-Fourth Annual IEEE Applied Power Electronics Conference and Exposition (APEC), Washington, DC, 2009, 739–743.

14. Tedeschi, E., Molinas, M., Carraro, M., and Mattavelli, P. Analysis of power extraction from irregular waves by all-electric power take off. IEEE Energy Conversion Congress and Exposition (ECCE), Atlanta, GA, 2010, 2370–2377.

15. Hazra, S., Shrivastav, A. S., Gujarati, A., and Bhattacharya, S. Dynamic emulation of oscillating wave energy converter. IEEE Energy Conversion Congress and Exposition (ECCE), Pittsburgh, PA, 2014, 1860–1865.

16. DATASHEET 160V MODULE. Maxwell Technologies, Inc., Global Headquarters, San Diego, CA, Document number: 3000246.6.

17. Casey, S. Modeling, simulation, and analysis of two hydraulic power take-off systems for wave energy conversion. Submitted for the degree of Master of Science in Mechanical Engineering, Oregon State University, 2013.

18. Engin, C. D., and Yeşildirek, A. Designing and modeling of a point absorber wave energy converter with hydraulic power take-off unit. 4th International Conference on Electric Power and Energy Conversion Systems (EPECS), Sharjah, United Arab Emirates, 2015.

19. Ruehl, K., Brekken, T. K. A., Bosma, B., and Paasch, R. Large-scale ocean wave energy plant modeling. IEEE Conference on Innovative Technologies for an Efficient and Reliable Electricity Supply (CITRES), Waltham, MA, 2010, 379–386.

20. Mueller, M. A. Electrical generators for direct drive wave energy converters. IEE Proceedings—Generation, Transmission and Distribution, 149(4), 446–456.

21. Polinder, H., Damen, M. E. C., and Gardner, F. Linear PM generator system for wave energy conversion in the AWS. IEEE Transactions on Energy Conversion, 19(3), 2004, September, 583–589.

22. Boström, C., Ekergård, B., Waters, R., Eriksson, M., and Leijon, M. Linear generator connected to a resonance-rectifier circuit. IEEE Journal of Oceanic Engineering, 38(2), 2013, April, 255–262.

23. Antonio de la, Villa-Jaen, Montoya-Andrade, D. E., and García-Santana, A. Control strategies for point absorbers considering linear generator copper losses and maximum excursion constraints. IEEE Transactions on Sustainable Energy, 9(1), 2018, January, 433–442.

24. Wu, F., Zhang, X.-P., Ju, P., and Sterling, M. J. H. Modeling and control of AWS-based wave energy conversion system integrated into power grid. IEEE Transactions on Power Systems, 23(3), 2008, August, 1196–1204.

25. Amundarain, M., Alberdi, M., Garrido, A. J., and Garrido I. Control strategies for OWC wave power plants. American Control Conference, Baltimore, MD, 2010, June.

26. M'zoughi, F., Bouallegue, S., Garrido, A. J., Garrido, I., and Ayadi, M. Stalling-free control strategies for oscillating-water-column-based wave power generation plants. IEEE Transactions on Energy Conversion, 33(1), 2018, March, 209–222.

27. Muthukumar, S., Kakumanu, S., Sriram, S., Desai, R., Babar, A. A. S., and Jayashankar, V. On minimizing the fluctuations in the power generated from a wave energy plant. IEEE International Conference on Electric Machines and Drives, San Antonio, TX, 2005, 178–185.

28. Ceballos, S., Rea, J., Lopez, I., Pou, J., Robles, E., and O'Sullivan, D. L. Efficiency optimization in low inertia wells turbine-oscillating water column devices. IEEE Transactions on Energy Conversion, 28(3), 2013, September, 553–564.

29. Songl, S.-K., and Park, J.-B. Modeling and control strategy of an oscillating water column-wave energy converter with an impulse turbine module. 15th International Conference on Control, Automation and Systems (ICCAS 2015), BEXCO, Busan, Korea, 2015, 1983–1988.

30. Takao, M., and Setoguchi, T. Air turbines for wave energy conversion. International Journal of rotating machinery. 2012, doi: 10.1155/2012/717398.

31. Corvelo, E. V. Analysis and Performance Comparison of an OWC Wave Power Plant Equipped with Wells and Impulse Turbines. 2014. https://fenix.tecnico.ulisboa.pt/downloadFile/395142739737/Resumo%20i4.pdf

32. Zobaa, A. F., and Bansal, R. C. (eds.). Handbook of Renewable Energy Technology. World Scientific Publishing Co. Pte. Ltd., Singapore, 2011.

33. Penche, C. Layman's Handbook on How to Develop a Small Hydro Site (Second Edition), European Commission, Bruselas, Belgica, 1998, June.

34. Munoz-Hernandez, G. A., Mansoor, S. P., and Jones, D. I. Modelling and Controlling Hydropower Plants, Springer-Verlag London, Dordrecht, Heidelberg, New York, 2013.

35. Lynn, P. A. Electricity from Wave and Tide, an Introduction to Marine Energy. John Wiley & Sons Ltd, Chichester, West Sussex, 2014.

36. Wang, L., Chen, S.-S., Lee, W.-J., and Chen, Z. Dynamic stability enhancement and power flow control of a hybrid wind and marine-current farm using SMES. IEEE Transaction Energy Conversion, 24(3), 2009, September, 626–639.

37. Wang, L., Lin, C.-Y., Wu, H.-Y., and Prokhorov, A. V. Stability analysis of a microgrid system with a hybrid offshore wind and ocean energy farm fed to a power grid through an HVDC link. IEEE Transactions on Industry Applications, 54(3), 2018, May/June, 2012–2022.

38. Valavi, M., and Nysveen, A. Variable-speed operation of hydropower plants. IEEE Industry Applications Magazine, 24(5), 2018, September/October, 18–27.

39. Steimer, P. K., Senturk, O., Aubert, S., and Linder, S. Converter-fed synchronous machine for pumped hydro storage plants. IEEE Energy Conversion Congress and Exposition (ECCE), Pittsburgh, PA, 2014, 4561–4567.
40. Argonne National Laboratory. Modeling and Analysis of Value of Advanced Pumped Storage Hydropower in the United States, 2014.
41. Kundur, P. Power System Stability and Control. McGraw-Hill, New York, 1994.
42. Liang, J., and Harley, R. G. Pumped storage hydro-plant models for system transient and long-term dynamic studies. IEEE PES General Meeting, Providence, RI, 2010, 1–8.
43. Hinkkanen, M., and Luomi, J. Modified integrator for voltage model flux estimation of induction motors. IEEE Transactions on Industrial Electronics, 50(4), 2003, August, 818–820.
44. Jayalath, S., and Hanif, M. Generalized LCL-filter design algorithm for grid-connected voltage-source inverter. IEEE Transactions on Industrial Electronics, 64(3), 2017, March, 1905–1915.
45. Janning, J., and Schwery, A. Next generation variable speed pump-storage power stations. 13th European Conference on Power Electronics and Applications, Barcelona, Spain, 2009, 1–10.
46. Pannatier, Y., Kawkabani, B., Nicolet, C., Schwery, A., and Simond, J.-J. Start-up and synchronization of a variable speed pump-turbine unit in pumping mode. XIX International Conference on Electrical Machines—ICEM, Rome, 2010, 1–6.
47. Krause, P. C., Wasynczuk, O., and Sudhoff, S. D. Analysis of Electric Machinery. IEEE Press, Piscataway, NJ, 2002.

6

Hybrid System Simulation

6.1 Plants with Diesel Generators

Despite the intensive development of new devices (fuel cells [FCs], microturbines) for the production of electricity in isolated and remote areas, diesel generators (DGs) remain the main source for generating electricity and/or as the backup devices for renewable energy system (RESs) with their variable power generation. Both synchronous generators (SGs) with a winding on a rotor as well as a permanent magnet synchronous generator (PMSG) and an induction generator (IG) can be used with the DG; the latter two types require the use of power electronics devices for proper functioning, whereas the SG allows direct connection to the network and performs voltage and frequency control.

The peculiarity of DG function is that its efficiency decreases significantly with decreasing produced power—it is considered that DG should work with a capacity of at least 20–30%. In this regard, the following options for functioning of DG are possible:

1. A DG is in operation continuously; when its power reduces to the minimum allowable level, either a ballast load is connected that prevents further power reduction or measures are taken to reduce the RES power. Specially developed resistive devices can be used as the ballast load (obviously, the least expedient solution) or some kind of payloads that can be switched on occasionally (heaters, pumps). One possible type of the ballast load may be compressor devices for storing high-pressure air. At those times when high power is required from the DG, this compressed air is used to pressurize the DG, which significantly reduces the need for fuel to realize the same power, Ref. [1]. Another possible solution of the problem of DG poor use at low loads is to use a DG with adjustable rotational speed; in this case, however, it is necessary to use a converter between the generator and the grid, Ref. [2].

   In such systems, the use of short-term energy storage devices (battery, supercapacitor, flywheel) is not mandatory, but they can be used to smooth out fluctuations of the frequency and load voltage during fluctuations in RES power, bearing in mind its intermittent nature and slow response of DG to disturbances, Ref. [3].

2. With a decrease in the DG-required power, it turns off and stops. In this case, the use of energy storage devices with medium or long storage time is mandatory. They must compensate for the possible power shortage in case of DG disconnection, as well as provide power to the load at the time of its start. In addition, they take part in maintaining the frequency and magnitude of the load voltage.

3. With a decrease in the DG-required power, its SG does not shut off, but the diesel engine ceases operation and stops; the SG disconnects from the diesel by the friction or electromagnetic clutch. In this case, the use of energy storage devices with medium or long storage time (the battery) is also mandatory. The SG is responsible for regulating the voltage of the load, and the battery for regulating its frequency, Ref. [4].

The first option, with the continuously operating DG, is simulated in the model **DieselH1**. A DG works together with a wind generator (WG) and photovoltaic (PV) cells. The diesel brings the SG into rotation with a power of 180 kVA at the voltage of 230 V. A diagram of the diesel engine model is shown in Figure 6.1, Ref. [5]. Note that the same diesel model is used in the previously described models **SG3b**, **SG4**. The SG excitation system controls the load voltage. The power of the WG is 200 kW at a base wind speed of 12 m/s, and the PV array is of 150 kW, their converter circuits consist of DC/DC boost converters, and inverters. The PMSG of a WG has 18 pairs of poles and a nominal rotation speed of 22 rad/s. The load current of the diode rectifier at their output is set as in the relationship (2.31), which in this case is written as

$$I_{ind}^{*} = (\omega_m^2 - D - F\omega_m)\,\omega_m\,\frac{P_n}{U_{rect}}. \tag{6.1}$$

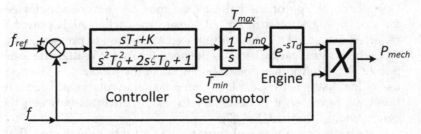

FIGURE 6.1
Diagram of the diesel engine model.

The appropriate circuits and units are placed in the subsystem **Opt**. The DC/DC converter of the PV array optimizes its output voltage, as was done in the model **Photo_9N**. The control systems for output inverters regulate their output currents to keep a predetermined voltage at the outputs of the above-mentioned DC/DC converters, which is assumed to be 450 V. The circuit applied in the model of the ballast load is taken from the model **Wind_DFIG_5**, Ref. [6], but its power is increased to 100 kW. The ballast load increases with reduction of diesel power to 0.2 P_{nom}.

The useful load consists of a permanent-connected part and two switched-off ones. The choice of power of the main (permanent) load has a great influence on the operation of the system. If the value of this power is large enough (approaching the maximum value of the RES power), then the actual power of RESs will not be enough to power the load most of the time, and the DG will be continuously in operation. If the load power is chosen to be significantly less than the maximal values of the RES power, then to ensure normal operation while increasing the actual power RES, it will be necessary to have a large ballast load (which is inefficient) or take other measures to reduce the generated RES power. These measures include controlling the angular position of the WG rotor blades and disabling a part of the parallel branches of the PV array. It is also possible to utilize reduction of the power produced by RESs by deviating from the optimal values of the rotational speed of the WG generator and of the PV output voltage. The power circuit diagram is depicted in Figure 6.2.

The process is shown in Figures 6.3 and 6.4 when, with the initial wind speed of 8 m/s, it begins to increase with the rate of 0.5 m/s² to the speed of 12 m/s, starting at $t = 3$ s, and then at $t = 17$ s it again decreases to 6 m/s. The relative irradiance of the PV array, initially equal to 0.25, begins to increase to 1 with the rate of 0.2 per 1 s at $t = 9$ s, and decreases to 0.4, starting at $t = 24$ s. Thus, there are times when both RESs generate a maximum

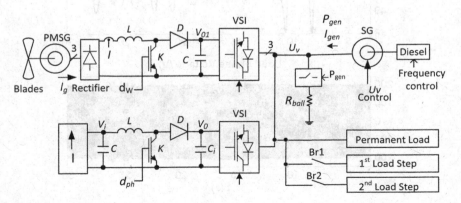

FIGURE 6.2
Block diagram of the microgrid with DG, WG, and PV array.

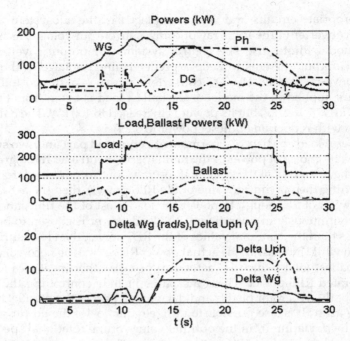

FIGURE 6.3
Processes in the model **DieselH1** under wind speed and illumination change.

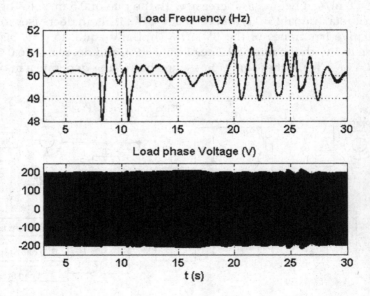

FIGURE 6.4
Load voltage and frequency change in the model **DieselH1**.

power of $200 + 150 + 0.2 \times 180 = 386$ kW (the last term is the minimum power of the DG). Of course, in reality, such an event occurs rarely, but, nevertheless, the system should be efficient even under such conditions. The power of the main load is 100 kW with the possibility of increasing in two steps by 50 kW each when the total theoretical power of the RESs is 170 kW and 220 kW (if these values can be estimated). Since even with the inclusion of all loads their total power is 300 kW, it is necessary to limit the generated power. This is achieved by generating signals *Delta Wg* and *Delta Uph*, causing a deviation from the optimal points. These signals are proportional to the existing RES power, increase with frequency of the load voltage increasing to 51 Hz, stop increasing when the frequency drops to 50.3 Hz, and decrease smoothly to zero with frequency decreasing to 49.5 Hz. It should be noted that the load powers are determined at a voltage of 220 V, whereas with the phase voltage fixed amplitude of 200 V, the effective value of the line voltage is given as $200 \times \sqrt{3}/\sqrt{2} = 245$ V, which leads to an increase in the load power relative to the nominal values by 24%.

It follows from Figures 6.3 and 6.4 that due to the effect of the signal *Delta Wg*, the WG power does not reach the maximum value of 200 kW. The power of the ballast load reaches 88 kW, and the power of the diesel engine does not decrease below a predetermined level. The frequency in transient conditions varies within ±1.5–2.0 Hz with the duration of the deviation being 1–2 s, and in the steady state is equal to the specified one. The load voltage is kept constant.

Figure 6.5 shows a more realistic process when the power of both RESs increases by turns—with the initial wind speed of 8 m/s, it begins to increase with the rate of 0.5 m/s^2 to the speed of 12 m/s, starting at $t = 3$ s and then again decreases to 6 m/s at $t = 9$ s. The relative irradiance of the PV array, initially equal to 0.25, at $t = 12$ s begins to increase to 1 with the rate of 0.2 1/s, and at $t = 18$ s decreases to 0.1. The main load is increased to 150 kW, and the *Delta Wg and Delta Uph* signals are deactivated. It can be seen that after reduction of the powers of both RESs, the load is supplied mainly from the DG, and the ballast load does not exceed 9 kW. The frequency deviations are the same as in the previous model.

Simulation in the model under consideration is extremely slow; at the same time, the time intervals at which the processes are studied must be long enough to consider possible combinations of modes. This phenomenon is typical when the hybrid systems are simulated. Further, the possible approaches to accelerate simulation are considered.

In the next few models, a version of the system with the SG permanently connected to the load is investigated. A friction clutch uniting it with a diesel engine can be modeled in various ways. In the toolbox *Simscape/SimDriveline* there is a model of such a clutch, but since this toolbox is not covered in the book, this model does not apply below. Three methods to simulate such a clutch can be suggested. In the first method, the processes of acceleration and

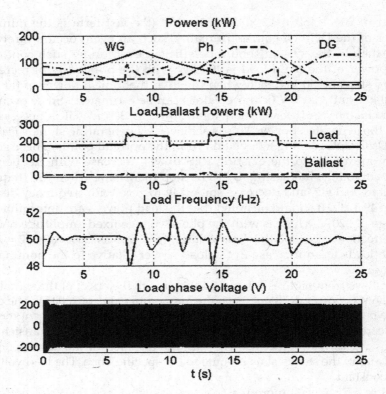

FIGURE 6.5
Processes in the model **DieselH1** under wind speed and illumination change in turns.

deceleration of a diesel engine are not simulated. When the clutch is disconnected, the torque applied to the SG is set to zero, and the diesel is simulated rotating with a nominal speed. When the clutch is turned on, the applied torque (power) of the SG is determined by the output torque (power) of the diesel engine. In the second method, the process of slowing down and accelerating a diesel engine is simulated. It is assumed that the rotational speeds of the diesel engine ω_d and the generator ω_g are governed by the following equations

$$\frac{d\omega_d}{dt} = \frac{T_d - T_{cl}}{J_d}, \qquad \frac{d\omega_g}{dt} = \frac{T_{cl} - T_g}{J_g}. \tag{6.2}$$

Here T_d is the torque of the diesel, T_g is the load SG torque, T_{cl} is the torque transferred by the clutch, and J_d, J_g are the moments of inertia of the rotating parts of the diesel engine and the generator, respectively. When the clutch is turned off, $T_{cl} = 0$, when it is turned on, $d\omega_d/dt = d\omega_g/dt$, from which

$$T_{cl} = \frac{T_d J_g + T_g J_d}{J_g + J_d}. \tag{6.3}$$

The second Equation (6.2) is included in the SG model, and the first should be added to the DG model. When the coupling is turned off, the fuel supply to the diesel engine stops, and it slows down to a complete stop, but this is impossible in the diesel model used previously, since its minimum torque is zero. To simulate the deceleration of a diesel engine, a braking torque equal to $a + b\omega_d$ is introduced into its model (e.g., in pu $a = 0.04$, $b = 0.1$). If it is necessary to put the diesel engine into operation, it accelerates, and at a speed close to the rotational speed of the SG, the *Clutch* signal is given to turn on the clutch.

To simulate clutch engagement dynamics, the following method can be used. The torque transferred by the clutch is written as

$$T_{cl} = C(\omega_d - \omega_g), \tag{6.4}$$

where $C = 0$ when the clutch is turned off, and C increases rapidly to a very large value $C \gg 0$ when the command to turn it on is received. Since the value of T_{cl} is limited (not more than 1.2 pu), the equality $\omega_d \approx \omega_g$ is implemented.

The first option is applied in the model **DieselH2**. The parameters of the DG and ballast load are the same as in the model **DieselH1**, the load power is 150 kW at the voltage of 220 V, the power of the WG is 350 kW, that is, it is equal to the total power of the WG and PV array in the previous model. The battery included in the model has a rated voltage of 490 V at a charging current of 260 A. It is connected to the AC microgrid using a voltage source inverter (VSI). The circuits for mode control are located in the subsystem **Mode_Control**.

The initial wind speed is 6 m/s, which provides a WG power of 42 kW; with the load power of 150 kW, the missing power is compensated with the DG. At $t = 5$ s, the wind speed increases to 12 m/s; at this, the power consumed from the generator drops. When the diesel power reduces to 0.2 of the nominal value, the trigger **Bistable** is reset, the battery is connected to the microgrid in a frequency control mode, and after 0.5 seconds, the diesel engine is disconnected from the generator, continuing to rotate at the nominal speed. Excess WG power over the load power is absorbed by the battery, operating in the charge mode; when the power of the battery reaches 95 kW, the ballast load is connected, limiting the charging current of the battery.

At $t = 17$ s, the wind speed begins to decrease to 6 m/s, and the power consumed by the battery drops. When it decreases to 75 kW, the power of the ballast load begins to decrease. The battery gradually goes into the discharge mode, maintaining the load power. When the power delivered by the battery reaches the value of 45 kW, the trigger **Bistable** is set, the battery is disconnected from the microgrid, and the diesel engine connects with the SG, providing the required load power. The obtained process is presented in Figure 6.6. It is seen that short-term deviations of frequency and load voltage are observed when the diesel engine connects to the SG.

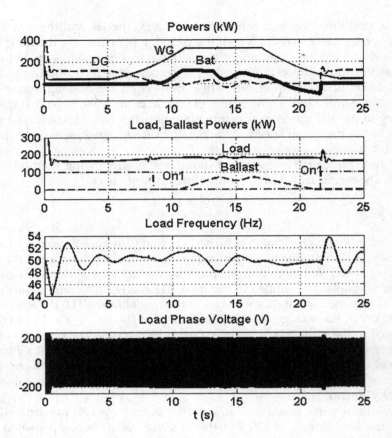

FIGURE 6.6
Processes in the model **DieselH2**.

The second option is used in the model **DieselH3**. With the signal *Run* = 1, the diesel engine accelerates to the nominal speed, and with *Run* = 0 it slows down to a full stop. The braking of a diesel engine is carried out with the help of the block of braking torque T_u, as mentioned above. The rotational speed of the diesel engine is calculated by the first Equation (6.2) at the output of the block **Integrator**. The torque transferred by the clutch is zero when the signal *Clutch* = 0 and is calculated by Equation (6.3) when *Clutch* = 1. The same torque rotates SG. The inertial constant of the generator was taken equal to 0.5 s, and the rotating parts of the diesel engine 0.7 s. When the diesel power diminishes to 0.2 of the nominal value, the triggers **Bistable** and **Bistable1** are reset, the diesel engine is disconnected from the SG and put into the braking mode; the battery is activated. When the wind speed decreases, the battery power reaches 45 kW in the discharge mode, the trigger **Bistable** is set, and the diesel engine starts to accelerate. When its rotational speed becomes equal to the rotational speed of the generator, the trigger **Bistable1**

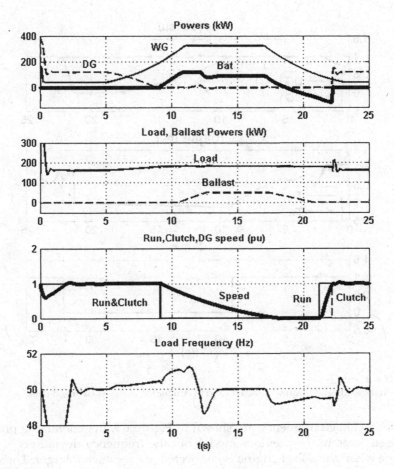

FIGURE 6.7
Processes in the model **DieselH3**.

is set, the signal *Clutch* = 1 is fabricated, which connects the diesel engine with the SG and turns off the battery.

The same scenario as for the previous model is shown in Figure 6.7. It is seen that the frequency deviations of the load voltage when the system devices are switched over are less than those in the previous model; the first axis of the scope **Load Voltage** shows that the voltage deviations are less too.

The model **DieselH4** is a modification of the previous one—the torque transferred by the clutch is determined not by Equation (6.3), but by Equation (6.4). When the *Clutch* command is given, value of C increases to a value of 10^4 with the speed of 10^4 1/s and when this command is disabled, value of C decreases to zero with the speed of 4×10^4 1/s. It can be seen in Figure 6.8 that when *Clutch* = 1, the rotational speeds of the diesel engine and the generator as well as their torques are the same. The values of powers

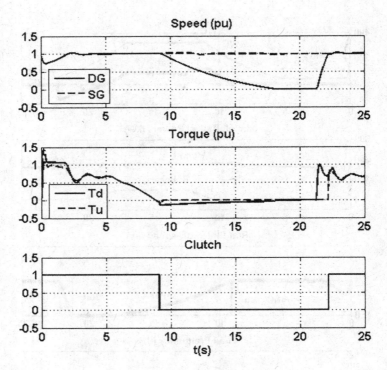

FIGURE 6.8
Rotational speeds and torques of the DG and SG under the clutch is switched over.

and load voltage frequency are shown in Figure 6.9. In general, the process
is the same as in the previous model, but the frequency deviations of load
voltage when the diesel engine is connected are somewhat larger. Thus, the
processes are to some extent determined by the characteristics of the coupling; for accurate modeling, its actual features must be taken into account.

It can be seen from the above discussion that when the actual power of RES
is large, the frequency is regulated by the battery with additional participation of the ballast load. However, it is possible that the battery is not able
to consume electricity (fully charged or faulty). In this case, the frequency
control function is transferred to the ballast load. However, since it cannot
generate electricity, the ballast load is put out of operation when the RES
power reduces, and it can no longer take part in frequency control. The only
way out is to turn on the diesel engine that can be done under reduction
of the ballast load to a certain minimum. In this case, however, some time
elapses between the detection of this event and the acceleration of the diesel
engine to the rated speed, during which the frequency of the voltage may
drop noticeably. The microgrid should withstand such conditions.

The described mode is investigated in the model **DieselH5**. Model parameters and the clutch control system are the same as in the model **DieselH4**.

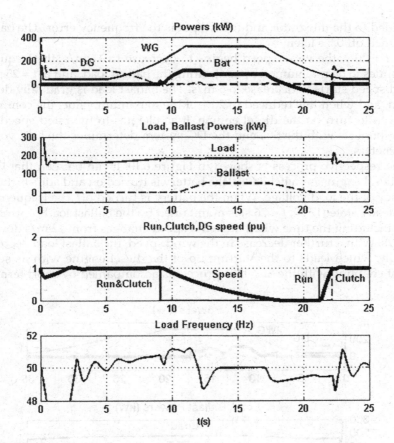

FIGURE 6.9
Processes in the model **DieselH4**.

Initially, the WG operates at a wind speed of 6 m/s (with a power of 42 kW), the diesel engine is connected with the SG and rotates at the nominal speed. At $t = 5$ s, the wind speed begins to increase to 12 m/s. At $t = 8$ s, the battery is turned on, and with a delay of 0.5 s, the diesel engine is disconnected from the SG and put into the braking mode. The grid voltage is maintained by the SG, and the frequency by the battery control system. When the battery power reaches 95 kW, the ballast load begins to increase. However, when $t = 16.5$ s, the command to switch off the battery is given. In this case, the function of frequency regulation goes to the ballast load control system, in which the proportional plus integral (PI) frequency controller is formed; the integrator of the ballast load controller is used as an integrator of this controller, and the P-circuit is created additionally. At first, the *B_off* command sets a new initial value for the integrator output—the value of $P_{bat}/600$ is added to its current value, where P_{bat} is the battery power before disconnection, and 600 W is the power of the ballast resistance step, then the parallel proportional circuit K_p

is added to the integrator and their input is the frequency error. The battery is turned off 0.5 s later.

At $t = 20$ s, the wind speed begins to decrease to 11 m/s, the frequency continues to be maintained by controlling the ballast load, and at $t = 25$ s, the wind speed starts to decrease to 6 m/s. The ballast load is gradually deactivated, and when it is reduced to 20% of the maximum value, the command is given to turn on the diesel engine. It accelerates to the rated speed and then connects with the SG, and the DG system determines the load voltage parameters.

The respective process is shown in Figure 6.10. It can be seen that when the diesel engine is switched off, the battery is turned on and adjusts the frequency of the load voltage. When the battery is turned off, the frequency in the steady state (17 s < t < 26 s) is maintained by the ballast load control system, including the time when the wind speed changes from 12 m/s down to 11 m/s. With a further decrease in the wind speed, the ballast load decreases rapidly, which leads to the starting up of the diesel engine with its subsequent coupling with the SG. This process is accompanied by short-term but

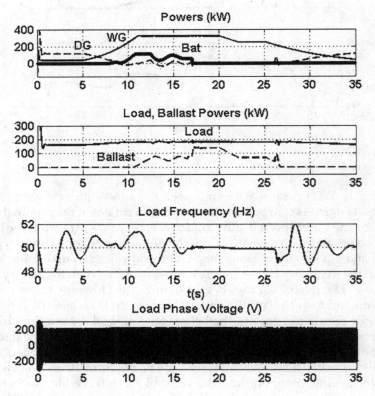

FIGURE 6.10
Processes in the model **DieselH5**.

noticeable frequency deviations; the magnitude of the frequency deviations depends on the rate of change of the wind speed and diesel engine acceleration; the load voltage is kept constant. With the course of time, the conditions of the devices in the microgrid and of the load are steadied; as the power of the WG decreases, the power of the DG increases.

The variant with generator disconnection is considered in the following model. In this case, it is necessary to decide on the structure of the control system. In the above-considered models, the SG remained connected to the AC microgrid, regardless of whether it was connected with a diesel engine or not, and its voltage played the role of a grid; the other devices synchronize their output voltages with the SG voltage. In the structure under consideration, the SG is switched off, and it is necessary for some device to be assigned to play the role of a grid. This may be a battery, but in this case, it should always be in operation, regardless of the need for it and the state of its charge, or assign this role to some RES device, for example, having the maximum power at the given time. The RES with sufficient power exists, because the situation is considered when RESs are producing sufficient power so that the diesel engine can be stopped.

Another possibility is to connect all the units with AC outputs in parallel, with utilization of the droop characteristics. This variant is employed further, since it is simpler to implement.

The WG, PV array, battery, and DG operate in the model **DieselH6**. The load power is 160 kW at the voltage of 220 V. The WG model has the power of 200 kW and is taken from the previous models, and its optimization is carried out by the relationship (6.1), but the control system of VSI-Gr is changed. Its output voltage is determined by the output of the **Three-Phase Sine Generator Sin_Ge** with the amplitude and frequency to be controlled. The frequency set-point is

$$f^* = f_0 - K_{fw} \frac{P_{el}}{P_{max}^* P_{nom}}, \tag{6.5}$$

where $f_0 = 50$ Hz, P_{el} is the power of the VSI-Gr output, and K_{fw} is the droop factor.

The wind turbine maximal mechanic power under wind speed V_w in pu to the nominal turbine power P_{nom} can be written as

$$P_{max}^* = (V_w / V_{wbase})^3 P_{base}^*,$$

where V_{wbase} is the wind base speed, and P_{base}^* is the maximal power at the wind base speed relating to the nominal power. **Sin_Ge** amplitude is determined by the controller of the VSI output voltage, the controller given value is

$$V^* = V_0 - K_{qw} Q_{el}, \tag{6.6}$$

where Q_{el} is the reactive power at the VSI output, K_{qw} is the droop factor, and V_0 is the load voltage with the required value $230 \times \sqrt{2}/\sqrt{3}$ V in the case. Besides, an action on the optimal rotational speed of the turbine is used in the control system of the DC/DC converter, with the purpose to reduce the produced power when the voltage at the VSI input (that is DC/DC converter output) is more than 550 V.

The model of the PV array with the power of 150 kW is adopted from the model **DieselH1**; the inverter control circuit is the same as for WG, with droop factors K_{fph}, K_{qph}. The battery model is the same as that used in previous models, but the magnitude and sign of the battery current are determined not by the frequency controller but by the load voltage controller. The DG model is also taken from the previous models, and the given values of the rotational speed and voltage of the generator are determined by relations similar to Equations (6.5) and (6.6), but in pu:

$$\omega^* = 1 - K_{\omega d} P_g, \quad V^* = V_0 - K_{vd} Q_g, \tag{6.7}$$

where P_g and Q_g are the active and reactive powers of the SG. Both signals are activated after connecting the SG to the grid. *LDG* inductance serves to reduce the current rush when the SG is connected to the microgrid, and a *RDG* resistor is needed, since a SG and an inductance cannot be connected in series to SimPowerSystems (the error message appears).

The circuits for connecting the SG to the microgrid are placed in the subsystem **DG_Control**. Simulation significantly slows down if the SG is idling or its load is not large enough. Therefore, before connecting the SG to the grid, it is loaded with 10 kW load via a three-phase switch that opens after the SG is connected to the grid. Naturally, in real conditions, this circuit is not required. The subsystem **Connection** generates the signal to connect the SG to the microgrid. When the model starts, both triggers **Bistable 1** and **Bistable 2** are reset, the SG is disconnected from the grid, and the diesel engine must, in principle, be stopped (in this model, to shorten the simulation time when the model starts, it rotates at a nominal speed, but before the start signal is given (*Run*) it slows down to about 70% of the rated speed due to lack of fuel supply).

If the RES total power exceeds the load power, the excess power is absorbed by the battery, which operates in the charge mode. When the RES power decreases, the battery goes into the discharge mode and compensates for the missing power keeping the load voltage. When the battery power in this mode increases to 45 kW, the signal *Run* is given to start the diesel engine. SG connects to the grid when the amplitude, frequency, and phase of its voltage are approximately equal to those of the microgrid. The frequencies and phases of the SG and microgrid voltages are measured by Phase Lock Loop (PLL) blocks. If the phases of these voltages are denoted by α and β, then $\sin(\alpha - \beta) \approx 0$ and $\cos(\alpha - \beta) \approx 1$ should be during synchronization. The calculations of sine and cosine are carried out in blocks **Subsin**, **Subcos**.

In principle, prior to connecting the SG to the grid, it is necessary to check the approximate equality of their voltages, however, this test is omitted in this model (it is assumed that this condition is met when the SG speed is close to the nominal one), since all units forming the microgrid have output voltage regulators with the same reference. When the frequencies and phases of the voltages being considered are close, the trigger **Bistable 2** is set, the signal *On* is generated, by which the SG connects to the grid, and after 0.5 s the battery is turned off. A number of oscilloscopes available in the subsystem are enabled to observe the synchronization process in detail and, if necessary, adjust the parameters.

The respective process is shown in Figure 6.11. Initially, the total power of both RESs is 240 kW and is spent on the load and for the battery charge. At $t = 2$ s, the irradiance of the PV array decreases from 0.8 to 0.3, and the wind speed decreases from 11 m/s to 8 m/s with the rate of 2 m/s^2. The power RESs is reduced, and the battery goes into the discharge mode. At $t = 2.8$ s, the battery power in this mode reaches 45 kW, and the signal is generated to

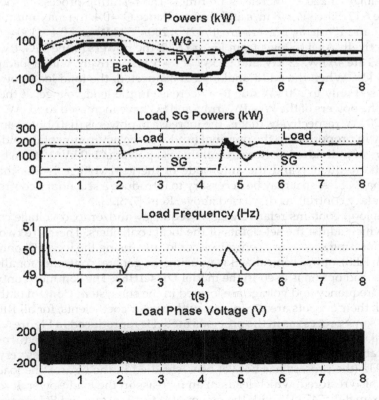

FIGURE 6.11
Processes in the model **DieselH6**.

start the diesel engine. The latter accelerates to the nominal speed, and after fulfillment of the synchronization conditions, the SG is connected to the microgrid, and then the battery is turned off. The SG power is approximately 100 kW and the rest of the power is provided by the RESs. Noticeable deviations of the load voltage at the time of synchronization are not observed. The frequency of the load voltage is slightly lower than the nominal value due to the influence of droop characteristics. It can be set equal to the nominal value by introducing a common secondary frequency controller, for which additional inputs df_{ref} are provided in the RES models, Ref. [7].

In the considered, rather complex system, various scenarios of the processes are possible; unfortunately, simulation in this system is extremely slow. Therefore, the possible options are not considered further, with the exception of the system response to the load increment.

It is assumed in the model **DieselH61** that the battery is not functioning; to speed up the simulation, its model has been removed. In addition, it is assumed that at $t = 0$ the signals *Run* and *On* already exist, the SG is connected to the microgrid. At $t = 5$ s, the load (whose value was 160 kW at the voltage of 220 V) increases 1.5 times. The resulting process is shown in Figure 6.12. Since $V_w = 8$ m/s and the irradiance $Q = 0.4$, the maximum values of WG and PV array powers are $(8/12)^3 \times 200 = 59$ kW and $0.4 \times 150 = 60$ kW, respectively. As it follows from Figure 6.12, the powers of the WG, PV array, and SG are 46 kW, 45 kW, and 90 kW at $t \approx 5$, respectively, with a load power of 180 kW. When the additional load is connected, the total load power has increased only to 230 kW due to a decrease in the load voltage; at the same time, the powers of the WG, PV array, and SG have increased to 55 kW, 54 kW, and 120 kW, respectively, that is, the increase in power is distributed approximately in proportion to the nominal (under given environmental conditions) powers of individual devices (recall that SG power is 180 kVA). It can be seen that the frequency and the amplitude of the grid voltage is less than the nominal values, so it may be necessary to introduce a secondary control loop (secondary control), as discussed above, Refs. [7, 8].

This loop contains relatively slow frequency and/or load voltage controllers, which adjust the set-points of the local controllers. There are two ways of performing secondary control: either locally (using the local PI controllers at each RES), or in a centralized way, using the general controller for all RESs. The second option is used in the model **DieselH62**. The common controllers of the frequency and voltage are located in the subsystem **Control** of the WG model; their outputs are the inputs, with proper coefficients, for all RESs.

The same process as in the previous model is reproduced in Figure 6.13, but at $t = 8$ s the secondary controllers of frequency and voltage are turned on. As it can be seen from this figure, the frequency restores to values very close to 50 Hz (the process has not yet been steadied in the figure), the load voltage is also restored, which leads to an increase in the load power. It reaches approximately 247 kW, with the power of WG, PV array, and SG increased to 57 kW, 55 kW, and 135 kW, respectively.

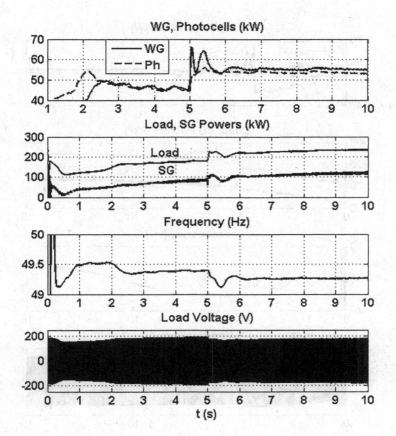

FIGURE 6.12
Processes in the model **DieselH61**.

6.2 Hybrid Systems with FC

Since even with the simultaneous use of different types of RESs, reliable supply of the autonomous load cannot be provided, in such sets, along with WG and PV arrays, sources of electricity are employed that can produce it regardless of the environmental conditions: DGs, FCs, and microturbines. In addition, the batteries are usually included in these installations with the purpose to smooth out transients when switching loads or individual power sources, as well as to provide power to critical loads during the start-up of a DG, FCs, or microturbines from the cold standby state. The installations with WGs, PV arrays, batteries, and FCs are considered next.

When the microgrid for the supply of an isolated load is designed, various combinations of the powers of its devices are possible, Refs. [9–11]. In the models considered below, the following assumptions are made: a three-phase

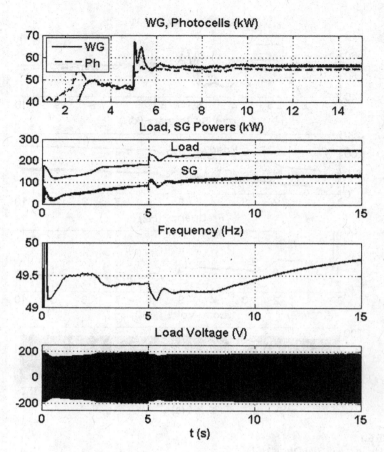

FIGURE 6.13
Processes in the model **DieselH62**.

load with the power of 100 kW at the voltage of 220 V consists of a permanent connected part with the power of 60 kW and a switched off (at low RES power) part with the power of 40 kW; WG power is 200 kW at the wind speed of 12 m/s (the WG model is taken from previous models with some changes); the power of the PV array is 150 kW under nominal environmental conditions (the optimization scheme is taken from the model **Photo_m01**); the battery has the rated voltage of 250 V and the discharge current of 260 A (the battery is equipped with a bidirectional boot/buck converter, as in model **SG3b**); the electrolyzer with a rated current of 750 A is described in Chapter 4; the PEMC FC unit consists of 250 cells with the area of 800 cm² each, having the rated voltage of 225 V at the rated current of 300 A.

There are two possibilities to construct the considered microgrid: to connect all individual devices in a DC link with the help of DC/DC converters,

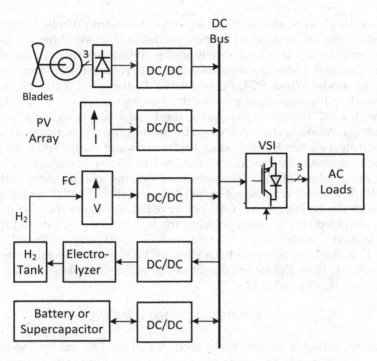

FIGURE 6.14
Block diagram of the microgrid with a common DC bus.

Figure 6.14, with a common inverter for power supply of the consumers, or with individual inverters for each device with their parallel connection to the consumer buses, as implemented in the model **DieselH6**, Figure 6.2. The first system is simpler because it contains fewer elements, while the second one requires synchronization of the individual inverters, but it is more flexible and allows the simple addition or removal of individual devices.

The first system is investigated in models **Wind_FC1**, **Wind_FC11**, **Wind_FC12**, in which various scenarios of processes are considered. The WG, PV array, and FC are connected to the DC bus via the boost converters DC/DC; to the electrolyzer via the buck converter, operating in the mode of output current regulation; and to the battery via a bidirectional boost/buck converter. The nominal DC voltage is assumed to be 450 V with the inverter output voltage of 220 V. For satisfactory operation of the microgrid, the adopted strategy of controlling its individual devices is very important. It is proposed in some works to use the battery charge level as a criterion for connecting and controlling devices; in this case, the battery must be permanently active. In other works, it is proposed to control the devices as a function of the relation between produced and consumed powers; at that, however, it is difficult to take into account additional power losses that can lead to mismatch between the powers and, as a consequence, to an uncontrolled increase or decrease in the DC voltage. In the next model, both possibilities are used.

Since simulation runs very slowly in the considered models, the devices that are disabled in the investigated process are removed from some models. So, in the model **Wind_FC1**, FC is deleted. In this model, the WG and PV array work in the maximum power extraction mode. Since the RES does not regulate the DC voltage in this mode, there must be a device that supports this voltage. As such, the battery can be used in the whole range, as well as the electrolyzer when RESs produce high power and the FC when the power produced is small. In this model, the load of 60 kW is connected permanently, and the load of 40 kW is connected when the total power of both RESs exceeds 105 kW, and is turned off when it decreases to 75 kW. The electrolyzer is connected when the excess total power of the RESs over the load power is 50 kW, provided that the total power of the RES exceeds 150 kW. In this case, the specified current of the electrolyzer is $I_{el}^* = (P_g - P_{load} - 5 \times 10^4)/200\,\text{A}$, where P_g is the total power of both RESs and 200 V is the electrolyzer operating voltage. With the rated current of 750 A, the complete loading of the electrolyzer is achieved under

$$P_g = 750 \times 0.2 + 50 + 100 = 300 \text{ kW.}$$

The battery switches to the charge mode when the DC voltage increases to 500 V and switches to the mode of energy transfer to the DC bus when this voltage decreases to 400 V. If the battery is not able to consume the amount of current necessary to maintain the DC voltage, for example, if it is fully charged, the electrolyzer takes over the voltage regulation, so that its current controller becomes interior for the voltage regulator. The processes under various scenarios are recorded in the following plots.

The process is depicted in Figure 6.15. With the initial wind speed of 5 m/s, it increases with the rate of 1 m/s^2, beginning at $t = 2$ s, to 12 m/s; the irradiance is equal to 0.33 and increases to 0.8 at $t = 11$ s. The DC voltage is maintained by increasing battery charge current. At $t \approx 5$ s, the second-stage load is turned on. The battery power decreases initially but afterwards increases again and reaches a maximum of 80 kW at $t \approx 7.8$ s when the electrolyzer starts. Thereafter, the battery power drops essentially, then increases again and reaches 85 kW at $t \approx 15$ s, whereas the electrolyzer power reaches 120 kW. Thus, there is an approximate balance (taking losses into account)—the produced power is 310 kW, the power consumption is 100 kW + 85 kW + 120 kW = 305 kW. The DC voltage and the load voltage are kept equal to the desired values with satisfactory accuracy.

Under certain weather conditions, for example in bright sunshine and strong winds, as well as in case of the emergency shutdown of some microgrid consumers, the generated power may exceed the allowable power of consumers, which will lead to an uncontrolled increase in the DC voltage. To eliminate this situation, it is possible either to connect the ballast load, or

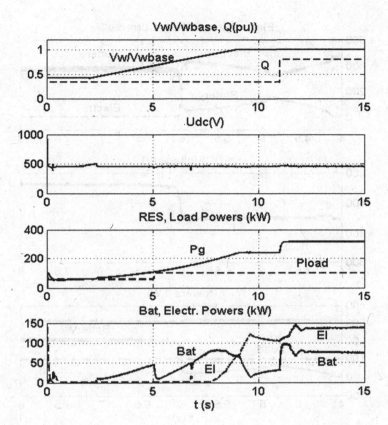

FIGURE 6.15
Processes in the model **Wind_FC1**.

in some way reduce the generated power. To implement the first option, it is necessary to have a controlled ballast load; its usage and the required power cannot always be predicted for sure, which increases the cost of installation. Since low-power WGs are rarely provided with devices for changing the position of the blades, to reduce the output power, the rotational speed of the generator can be increased as compared to the optimal one. To reduce the power produced by the PV array, it is possible to increase its output voltage compared with the optimal value. In the model **Wind_FC11**, when U_{dc} increases to 520 V, smoothly increasing signals dW and dU are formed, which cause the above-mentioned changes in speed and voltage. Decreasing the DC voltage to 460 V reduces these signals to zero. In Figures 6.16 and 6.17, processes are displayed as that at a wind speed of 12 m/s and degree of irradiation 0.8 ($P_g = 200 + 150 \times 0.8 = 320$ kW), the electrolyzer is turned off at $t = 6$ s, and all excess power of 220 kW must be created by the battery that exceeds its permissible parameters. This process is reproduced in Figure 6.16. It can be seen that, despite the increase in the AC current to

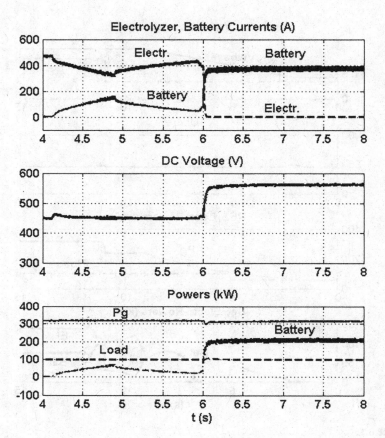

FIGURE 6.16
Processes in the model **Wind_FC11** without RES power limitation.

400 A, the DC voltage is set at 560 V. With the introduction of the above-described correcting feedback, the generated power decreases to 220 kW in this situation, so that the AC power should be only 120 kW. DC voltage is set at 480 V (Figure 6.17).

Model **Wind_FC12** simulates the process when a task to control DC voltage, when the RES power is sufficiently great, switches over from the battery to the electrolyzer; for example, if the battery is fully charged. With the initial wind speed of 6 m/s and the relative irradiance of 0.8, the first value increases with the rate of 1 m/s², starting at $t = 2$ s. At $t = 5$ s, the decision is made to disconnect the battery. At that, the control structure of the electrolyzer changes—its current controller becomes interior for the voltage controller. The battery is turned off at $t \approx 7$ s. Figure 6.18 depicts the process when setting the maximum current value of the electrolyzer (the PI controller in the subsystem **El_Contr**) equal to 1000 A. It can be seen that the DC voltage is kept with sufficient accuracy. In Figure 6.19, the maximum current

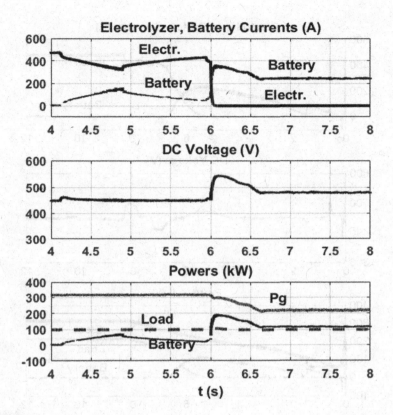

FIGURE 6.17
Processes in the model **Wind_FC11** with RES power limitation.

of the electrolyzer is 800 A. Now its power is insufficient, and the previously mentioned circuits for the decrease of the produced power come into effect. DC voltage is limited to 520 V.

The model **Wind_FC13** simulates the process of enabling FC when the RES power decreases. The wind speed, initially equal to 10 m/s, decreases with the rate of 1 m/s², beginning at $t = 2$ s, to the speed of 5 m/s. At $t = 4$ s, the irradiance, initially equal to 0.8, decreases to the value of 0.2 with the rate of 0.2 1/s. The process is reproduced in Figure 6.20. At $t = 5.13$ s, the electrolyzer is turned off, the DC voltage is maintained by the FC. The switchable load stage is switched off at $t = 6.4$ s. The power load 60 kW is met by RES (40 kW) and FC (20 kW).

It should be noted that the control algorithms used above are simplified. In fact, they should take into account the real state of the battery and the hydrogen storage tank, optimize the operation of the simultaneously functioning devices, taking into account possible changes of their parameters, etc. These algorithms depend on the power of the devices in the microgrid, the load

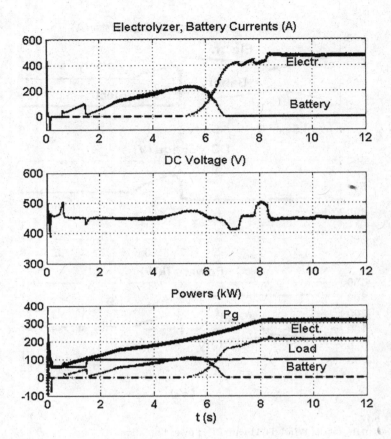

FIGURE 6.18
Processes in the model **Wind_FC12** with electrolyzer maximum current 1000 A.

requirements, climatic conditions, etc., and are quite complex in the general case; simulation of processes with such algorithms takes a long time. Those readers who are engaged in the development of such algorithms should do this work; taking into account the requirements mentioned above, they can use the models developed above as an initial system.

In the subsequent models, the hybrid system with a common AC bus is investigated. The principles of control of such systems were investigated in the previous section—the use of droop characteristics and, if necessary, the use of a secondary frequency and voltage control loops. Since the addition of inverters slows down the simulation even more, the simulation in these models is carried out fragmentarily. WG and PV array models are the same as in the model **DieselH6** (200 kW at the base wind speed of 12 m/s and 150 kW at nominal conditions, respectively).

The case is simulated in the model **Wind_FC2** (**Wind_FC2N** for R2016b, R2019a) when the RES power is large enough so that the electrolyzer can be

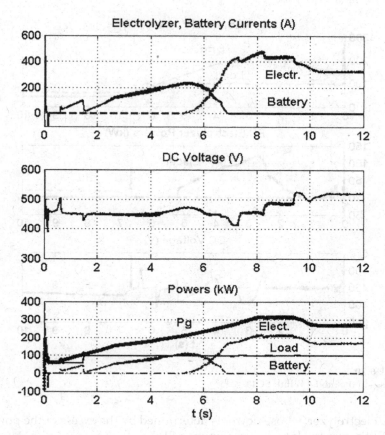

FIGURE 6.19
Processes in the model **Wind_FC12** with electrolyzer maximum current 800 A.

activated. In previous models, a DC/DC buck converter was used for connection with the DC power source. Since the transfer of energy in the electrolyzer is unidirectional, a diode rectifier could be used to power this converter in this configuration. However, its input current is highly distorted, and since the power of the electrolyzer is a noticeable amount in the total power of the microgrid, the voltage produced by the microgrid would be also greatly distorted. Therefore, a controlled rectifier is used in the model, as in the battery model. The number of the electrolyzer cells is doubled with the reduction of area of each cell.

The load power is 200 kW in this model. When $t < 2$ s, the battery operates, keeping up the desired value of the load voltage. The total power of both RESs is consumed for powering the load and recharging the battery. At $t = 2$ s, the command to turn on the electrolyzer is given. The battery goes into the discharge mode, producing the additional power to supply the electrolyzer. At $t = 2.5$ s, the battery is turned off, and the load voltage is maintained

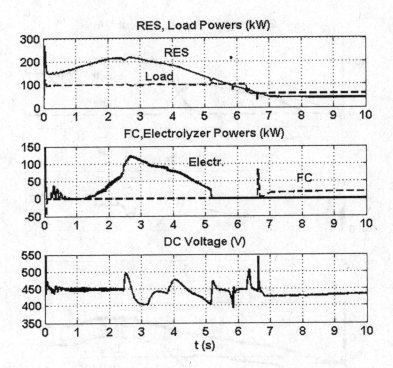

FIGURE 6.20
Processes in the model **Wind_FC13**.

by the electrolyzer, whose power is determined by the excess of the power of both RESs over the load power (Figure 6.21).

The model **Wind_FC21** (**Wind_FC21N** for R2016b, R2019a) simulates operation when the RES power is less than the load power, which is 60 kW at a nominal voltage of 220 V. The missing power is first provided by the battery, which keeps up the specified load voltage, and then by the FC, which takes over the battery functions, and the latter is turned off. In the FC models, 550 cells are connected in series; FC power is estimated as 60 kW. The FC output voltage is the DC voltage of the inverter (an intermediate DC/DC converter is not used). The following schedule is implemented in the model: the power of the WG is 19 kW, the power of the PV array is 20 kW (without taking into account the influence of the droop and losses); the battery provides the power of about 20 kW. At $t = 0.5$ s, preparation begins for FC switching; it is activated and connected to the grid with zero current demand. At $t = 1.5$ s, the current controller of the FC inverter (I_d component) receives the assignment equal to

$$I^*_{fcd} = \sqrt{2}(P_{load} - P_g)/220/\sqrt{3}\ A;$$

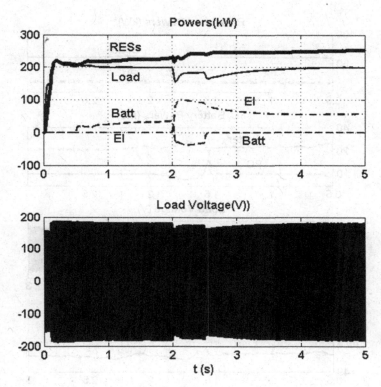

FIGURE 6.21
Processes in the model **Wind_FC2**.

which is sufficient to relieve the battery the latter, nevertheless, remains in the operation, controlling the load voltage. At $t = 2$ s, the battery is turned off, the FC inverter begins to control the load voltage, having the interior current controller. As it can be seen from Figure 6.22, the load voltage is kept with reasonable accuracy.

6.3 Microgrid Simplified Simulation

In most cases, a microgrid often contains quite a large number of generators driven by diesel or microturbine and RESs. Complete simulation of such systems takes a long time, since the RESs contain the power electronics devices that require a small, no more than 10 ms, calculation sampling time, while the process must be simulated for at least several minutes. When

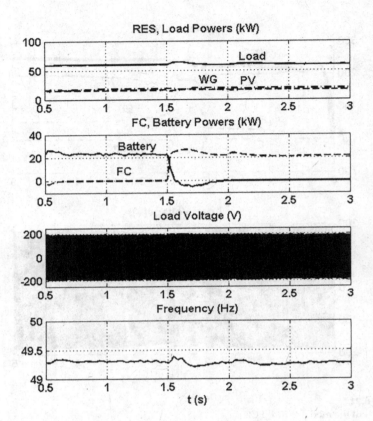

FIGURE 6.22
Processes in the model **Wind_FC21**.

these systems are investigated, it is often necessary to determine quickly the certain characteristics of the microgrid, which determine its overall quality. These characteristics include changes in the microgrid frequency and voltage in the event of a discrepancy between the generated and consumed power, resulting from an abrupt change in the load or switching off the generating devices. These issues are discussed in Ref. [12], on whose ideas further reasoning is based.

The main parameter characterizing the stability of the microgrid is the frequency change that occurs when there is the abrupt mismatch between the generated and consumed power; at that, the voltage changes are less important because they are successfully and quickly counteracted by the regulators of the generator excitation systems. Thus, it is desirable to be able to quickly estimate the process of changing the frequency under the above-mentioned disturbances.

The synchronous generators, their inertial masses, and primary engine control systems (turbine-governor systems) play the main role in the dynamics of frequency change. Therefore, it seems reasonable to replace the entire microgrid with one equivalent synchronous generator with its primary mover, whose response to load disturbance will be close to the response of the entire network under the same disturbance. When such a generator is designed, the following assumptions are made.

1. Synchronous generators are connected to common buses directly, without power electronics devices (DGs, microturbines with separate axes). It should be kept in mind that, to this day, DGs remain the primary sources in the microgrid, ensuring its operation under insufficient RES power (see, for example, the description of the autonomous system on King Island in North West Tasmania, Ref. [13]).

2. Since all the energy sources are located close to each other in the autonomous system, it is possible to neglect the resistances of the connecting wires and take all the sources connected in parallel.

3. Equivalent generator power is equal to

$$S = \sum_{i=1}^{n_g} M_i + \sum_{j=1}^{n_r} S_j, \qquad (6.8)$$

where n_g is the number of the real generators, M_i are their rated powers, n_r is the number of RESs, and S_j are their powers under given environmental conditions.

4. RESs have zero inertia constant, if only a virtual inertia is not implemented in their control system (Chapter 2). Then the inertia constant of the equivalent generator is defined as

$$H = \sum_{i=1}^{n_g} \frac{H_i M_i}{S}, \qquad (6.9)$$

where H_i are the generator inertia constants, including the inertia of the primary movers.

5. Parameters of the equivalent circuit of the equivalent generator in pu are equal to the weighted sum, with weights $M_i / \sum_{j=1}^{n_g} M_j$, of the same parameters of the existing generators.

6. Since RES powers are determined by external conditions and cannot be regulated, the transfer function of the governor-diesel system is defined as

$$W = \sum_{i=1}^{n_g} \frac{W_i M_i}{S}, \qquad (6.10)$$

where W_i are the transfer functions of the governor-diesel systems of each DG.

The considered method is investigated in the model **DieselS1**. Three DGs have the powers of 100 kW, 70 kW, and 50 kW at a voltage of 230 V. Their inertial constants are 0.7 s, 0.6 s, and 0.5 s, respectively. For simplicity, their parameters in pu and control systems are the same. The generators are connected in parallel to the load together with the WG, whose power is 350 kW at the wind speed of 12 m/s. The wind speed is 10 m/s under simulation, so the power of the WG is $350 \times (10/12)^3 = 202.5$ kW. Thus, the capacities of traditional and renewable sources are approximately equal. The load of 200 kW increases to 400 kW at $t = 5$ s. Parameters of the equivalent generator are: power 422 kW and inertia constant $H = (0.7 \times 100 + 0.6 \times 70 + 0.5 \times 50)/422 = 0.325$ s; since all control systems are the same, the same structure of the control system is adopted for the equivalent generator, but its effect is reduced in relation to 220/422. Figure 6.23 shows amplitude and frequency variations in the load voltage for the complete and equivalent systems. It can be seen that they do not differ significantly. Thus, using an equivalent generator, one can quickly

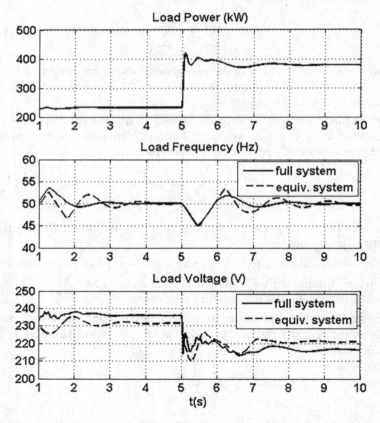

FIGURE 6.23
Comparison of the complete and simplified microgrid model.

estimate the main characteristics of the system, determine whether a change in the composition and parameters of energy sources is required, and determine protection settings. At the same time, in the equivalent model there are no elements of power electronics, which allows increasing the calculation sample time to 50 ms that accelerates the simulation by about 15 times; the difference will increase with the complexity of the microgrid.

As already mentioned, this simplified model can be used for investigation of the methods to improve the quality of the microgrid; in this case, to reduce the frequency of deviation when the load changes. In the model **DieselS2**, a battery is used for this purpose, the control circuit of which is taken from the model **DieselH4**. Of course, the calculation process slows down, especially since employment of the inverter requires reduction of the sampling time, but it runs anyway faster in this simplified model. Figure 6.24 shows the processes in this model. It is seen as the diesel gradually takes the full load; the maximum frequency deviation decreased from 5 Hz to 2 Hz, in 2–2.5 s the frequency restores.

In the model **DieselS3**, the battery from the previous model has been added to the full model **DieselS1**. After carrying out simulation it can be seen that the frequency change is almost the same as shown in Figure 6.24, but

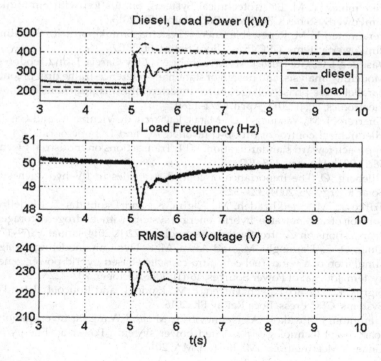

FIGURE 6.24
Process in the model **DieselS2**.

the simulation is performed four times slower than in the model **DieselS2**. As mentioned above, the difference will increase with complication of the hybrid system.

References

1. Ibrahimab, H., Ilincaa, A., Younesc, R., Perronb, J., and Basbousa, T. Study of a hybrid wind-diesel system with compressed air energy storage. IEEE Canada Electrical Power Conference, Montreal, Quebec, Canada, 2007, 320–325.
2. Chen, Z., and Hu, Y. A hybrid generation system using variable speed wind turbines and diesel units. The 29th Annual Conference of the IEEE Industrial Electronics Society IECON'03, Roanoke, VA, 2003, 2729–2734.
3. Tankari, M. A., Camara, M. B., Dakyo, B., and Nichita, C. Attenuation of power fluctuations in wind diesel hybrid system—Using ultracapacitors and batteries. XIX International Conference on Electrical Machines—ICEM, Rome, 2010, 1–6.
4. Sebastián, R., and Peña, R. Simulation of a high penetration wind diesel system with a Ni-Cd battery energy storage. 35th Annual Conference of IEEE Industrial Electronics IECON'09, Porto, Portugal, 2009, 4589–4594.
5. Perelmuter, V. M. Electrotechnical Systems, Simulation with Simulink® and SimPowerSystems. CRC Press, Boca Raton, FL, 2013.
6. Perelmuter, V. M. Renewable Energy Systems, Simulation with Simulink and SimPowerSystems. CRC Press, Boca Raton, FL, 2017.
7. Vasquez, J. C., Guerrero, J. M., Savaghebi, M., Eloy-Garcia, J., and Teodorescu, R. Modeling, analysis, and design of stationary-reference-frame droop-controlled parallel three-phase voltage source inverters. IEEE Transactions on Industrial Electronics, 60(4), 2013, April, 1271–1280.
8. Guerrero, J. M., Vasquez, J. C., Matas, J., García de Vicuña, L., and Castilla, M. Hierarchical control of droop-controlled AC and DC microgrids—A general approach toward standardization. IEEE Transactions on Industrial Electronics, 58(1), 2011, January, 158–172.
9. Ulleberg, Ø. The importance of control strategies in PV–hydrogen systems. Solar Energy, 76, 2004, 323–329.
10. Trifkovic, M., Sheikhzadeh, M., Nigim, K., and Daoutidis, P. Modeling and control of a renewable hybrid energy system with hydrogen storage. IEEE Transactions on Control Systems Technology, 22(1), 2014, January, 169–179.
11. Onar, O. C., Uzunoglu, M., and Alam, M. S. Dynamic modeling, design and simulation of a wind/fuel cell/ultra-capacitor-based hybrid power generation system. Journal of Power Sources, 161(1), 2006, October, 707–722.
12. Sigrist, L., Lobato, E., Echavarren, F. M., Egido, I., and Rouco, L. Island Power Systems. CRC Press, Boca Raton, FL, 2016.
13. Hamilton, J., Tavakoli, A., Negnevitsky, M., and Wang. X. Investigation of no load diesel technology in isolated power systems. Power ad Energy Society General Meeting (PESGM), Boston, MA, 2016, 1–5.

List of the Appended Models

DieselH1: WG, Photo array and permanently operating DG with isolated load

DieselH2: WG, Battery and DG with friction coupling, 1st variant

DieselH3: WG, Battery and DG with friction coupling, 2nd variant

DieselH4: WG, Battery and DG with friction coupling, 3rd variant

DieselH5: WG, Battery and DG with friction coupling, with temporary disconnected Battery

DieselH6: DG, WG, Photocells and Battery, DG start

DieselH61: DG, WG, Photocells, Load Change

DieselH62: DG, WG, Photocells, Load Change, with secondary Control

DieselS1: WG and 3 DG with isolated load; comparison of the detailed and simplified models

DieselS2: Microgrid equivalent model with Battery

DieselS3: WG, 3 DG and Battery, detailed model

Electr1: PEMFC and Electrolyzer

Fuel_1: PEMFC model

Fuel_sof: SOFC model

Fuel_sof1: SOFC connected to Three-Phase Grid

Fuel2: Fuel cell and Ultracapacitor, fuel interruption

HYDRO_PMSG_1: Small Hydraulic Turbine with PMSG, sensorless control

HYDRO_PMSG_1a: Small Hydraulic Turbine with PMSG, sensorless control, simplified model

HYDRO_PMSG_2: Small Hydraulic Turbine with PMSG, P&O method

HYDRO_PMSG_3: Small Hydraulic Turbine with PMSG, P&O method, I_q tuning

HYDRO_PMSG_4: Small Hydraulic Turbine with PMSG and DC/DC Converter, P&O method

IG_6PhN: Diesel and 6-Phase IG

M2CC: MMC, circulation current suppression

MTURBO1M: Microturbine, Power Electronics and Grid

MTURBO2M: Microturbine, Power Electronics and Load

MTURBO3M: Microturbine with Speed Control (sensorless model)

MTURBO4M: Microturbine start, simplified turbine model

MTURBO4M1: Microturbine start, detailed turbine model

MTURBO5M: Microturbine start with isolated load

MTURBO6M: Microturbine with diode rectifier and grid

MTURBO61M: Microturbine with diode rectifier and isolated load

MTURBO7M: WG and Microturbine with isolated load

MTURBO7M1: WG and Microturbine with isolated load, WG start

MTURBO8M: FC and Microturbine working jointly, DC load

MTURBO81M: FC and Microturbine working jointly, AC load

MTURBO81MN: FC and Microturbine working jointly, AC load, R2016, R2019 versions

OWC1: Simple OWC model, 1st version

OWC1a: Simple OWC model, 2st version

OWC1b: OWC with IG

OWC2: OWC with DFIG, detailed model

OWC2a: OWC with DFIG, simplified model

OWC2c: OWC with DFIG, optimal torque control

OWC2b: OWC with DFIG, optimal speed control

OWC2d: OWC with DFIG, flow coefficient control

OWC2e: OWC with DFIG, VSI-Ge and VSI-Gr, flow coefficient control

OWC3c: OWC with Impulse Turbine and PMSG

OWC3d: Four OWCs with Impulse Turbines and PMSGs

PA1: Point Absorber with AC output

PA2: Point Absorber with DC output

PA3c: Point Absorber and AC grid

PA3: Three Point Absorbers in parallel and AC grid

PA3b: Three Point Absorbers in parallel and AC grid non-monochromatic wave

PA4b: Three Point Absorbers with VSI-Ge in parallel and AC grid

PA32: Three Point Absorbers in parallel, UC and AC grid

PA5: Point Absorber with VSI-Ge and Magnetic Lead Screw

PA6s: Model of the Hydraulic Power Take-off System

PA6: Point Absorber with the Hydraulic Power Take-off System

PA6b: Point Absorber with the Hydraulic Power Take-off System and AC grid

PA7: Point Absorber with virtual linear PMSG and AC load

PA7b: Point Absorber with virtual linear PMSG and AC grid

Photo_9N: Parallel connection of Ph-cells in the power PV Array, common Grid VSI

Photo_9N_N: Parallel connection of Ph-cells in the power PV Array, common Grid VSI, R2016b, R2019a versions

Photo_9N1: Parallel connection of Ph-cells in the power PV Array, PV model from SimPowerSystems, common Grid VSI

Photo_9N1_N: Parallel connection of Ph-cells in the power PV Array, PV model from SimPowerSystems, common Grid VSI, R2016b, R2019a versions

Photo_9N2: Parallel connection of Ph-cells in the power PV Array, individual Grid VSI

Photo_9N3: Parallel connection of Ph-cells in the power PV Array, PV model from SimPowerSystems, individual Grid VSI

Photo_9N4: Parallel connection of Ph-cells in the power PV Array, common Grid VSI, grid fault

Photo_9N4_N: Parallel connection of Ph-cells in the power PV Array, common Grid VSI, grid fault, R2016b, R2019a versions

Photo_Hinv1: PV_Arrays with Multilevel Cascaded H-bridge Inverter

Photo_Hinv1_2015a: PV_Arrays with Multilevel Cascaded H-bridge Inverter, SimPowerSystems PV model

Photo_Hinv2a: PV_Arrays with Multilevel Cascaded Inverter, with common transformer 1000 Hz

Photo_Hinv2a_2015a: PV_Arrays with Multilevel Cascaded Inverter, with common transformer 1000 Hz, SimPowerSystems PV model

Photo_Hinv1N_2016: PV_Arrays with Multilevel Cascaded H-bridge Inverter, for R2016b, R2019a.

Photo_Hinv1NN: PV_Arrays with Multilevel Cascaded H-bridge Inverter, **Full-Bridge MMC**

Photo_Hinv2aN: PV_Arrays with Multilevel Cascaded Inverter, with common transformer 1000 Hz, for R2016b, R2019a

Photo_Hinv2aNN: PV_Arrays with Multilevel Cascaded Inverter, with common transformer 1000 Hz, **Full-Bridge MMC**

Photo_m01: PV and single-phase grid working parallel and separately, island mode detection

Photo_m02: PV and three-phase grid working parallel and separately, island mode detection, two measuring PVs

Photo_m02_2015a: PV and three-phase grid working parallel and separately, island mode detection, two measuring PVs, PV model from SimPowerSystems

Photo_m02a: PV and three-phase grid working parallel and separately, one measuring PV

Photo_m02a_2015a: PV and three-phase grid working parallel and separately, one measuring PV, PV model from SimPowerSystems

Photo_m02aN: PV and three-phase grid working parallel and separately, one measuring PV, R2016b, R2019a versions

Photo_m02aN_2015a: PV and three-phase grid working parallel and separately, one measuring PV, PV model from SimPowerSystems, R2016b, R2019a versions

Photo_m02N: PV and three-phase grid working parallel and separately, island mode detection, two measuring PVs, R2016b, R2019a versions

Photo_m02N_2015a: PV and three-phase grid working parallel and separately, island mode detection, two measuring PVs, PV model from SimPowerSystems, R2016b, R2019a versions

Photo_m03: Two PV Arrays in the island mode

Photo_m03_2015a: Two PV Arrays in the island mode, PV model from SimPowerSystems

Photo_m1: Characteristics of the author's PV sell model

Photo_m1N: Characteristics of the PV string model of SimPowerSystems.

PMSG_6phN: 6-Phase PMSG and resistive load

Pump1: Pumped Storage Hydro-Plant with constant speed SG
Pump2: Pumped Storage Hydro-Plant with variable speed SG, simplified model
Pump21: Pumped Storage Hydro-Plant with variable speed SG and 3-level VSIs, simplified model
Pump22: Pumped Storage Hydro-Plant with variable speed SG and 3-level VSIs
Pump221: Pumped Storage Hydro-Plant with variable speed SG and 3-level VSIs, bypass simulation
Pump222: Pumped Storage Hydro-Plant with variable speed SG and 3-level VSIs, with Grid Filter
Pump23: Pumped Storage Hydro-Plant with DFIG, start simulation
Pump24: Pump/Turbine with DFIG, steady-state, simplified model
Pump242: Pump/Turbine with DFIG, steady-state, detailed model
Pump2N: Pumped Storage Hydro-Plant with variable speed SG, simplified model for R2019a
Pump21N: Pumped Storage Hydro-Plant with variable speed SG and 3-level VSIs, simplified model for R2019a
Pump22N: Pumped Storage Hydro-Plant with variable speed SG and 3-level VSIs for R2019a
Pump221N: Pumped Storage Hydro-Plant with variable speed SG and 3-level VSIs, bypass simulation, for R2019a
Pump222N: Pumped Storage Hydro-Plant with variable speed SG and 3-level VSIs, with Grid Filter, for R2019a
SG_6PhN: Diesel Engine and 6-Phase SG
SG_9ph: 9-Phase SG and resistive load
SG_9phN: 9-Phase SG and resistive load, 2nd version
SG3b: Frequency Support with Battery
SG4: Frequency Support without Battery
Synchrover1: Synchronverter model
Synchrover2: Wind-Generator and synchronverter
Tide_mod1: Tidal Power Plant
Tide_mod2: Tidal Power Plant with power limiting
Tide_mod3a: Tidal Power Plant, turbine start
Tide_mod6a: Tidal Stream Turbine with PMSG, simplified model
Tide_mod6: Tidal Stream Turbine with PMSG
Tower: Subsystem for modeling Tower Shadow
Wind_DFIG_1N: WG with Wounded Rotor IG, Tower Shadow and Drive Train
Wind_FC1: WG, PV, Battery and Electrolyzer, DC Bus
Wind_FC11: WG, PV, Battery and Electrolyzer, DC Bus, RES power limitation
Wind_FC12: WG, PV, Battery and Electrolyzer, DC Bus, Battery is disconnecting
Wind_FC13: WG, PV, Battery and Electrolyzer, DC Bus, FC is connecting
Wind_FC2: WG, PV, Battery and Electrolizer with isolated load

Wind_FC2N: WG, PV, Battery and Electrolizer with isolated load, for R2016b, R2019a

Wind_FC21: WG, PV, Battery and FC with isolated load

Wind_FC21N: WG, PV, Battery and FC with isolated load, for R2016b, R2019a

Wind_IG_6N: Wind Turbine with 6-Phase IG

Wind_PMSG_1N: WG with PMSG and power grid, power train simulation

Wind_PMSG_1Na: WG with PMSG and power grid, direct-power control

Wind_PMSG_4N: 6-Phase PMSG and power grid

Wind_PMSG_8N: Off-Shore Park with two terminals, with priority

Wind_PMSG_8Na: Off-Shore Park with two terminals, with proportional sharing

Wind_PMSG_9M: Off-Shore Park with series DC connected WG, MMC inverter

Wind_PMSG_9M1: Off-Shore Park with series DC connected WG with high-frequency transformer, CSI inverter

Wind_PMSG_9Ma: Off-Shore Park with series DC connected WG, MMC inverter, with the block **Half-Bridge MMC**

Wind_PMSG_9N: Off-Shore Park with series DC connected WG, 3-level inverter

Wind_PMSG_9N1: Off-Shore Park with series DC connected WG without transformer, CSI inverter

Wind_SG_2aN: Wind—6-Phase Synchronous Generator with Drive Train

Wind_SG_4: Wind—9-phase SG

Wind_SG_4a: Wind—9-phase SG, high harmonics consideration

Wind_SG_5: Wind—9-phase SG, 2nd version

Wind_SG_5a: Wind—9-phase SG, 2nd version, high harmonics consideration

Index

Page numbers followed by *f* and *t*
indicate figures and tables, respectively

Printed in the United States
by Baker & Taylor Publisher Services

Printed in the United States
by Baker & Taylor Publisher Services